All That Is

All That Is

A Naturalistic Faith for the Twenty-First Century

Arthur Peacocke

*A theological proposal
with responses from leading thinkers
in the religion-science dialogue*

Edited by Philip Clayton

FORTRESS PRESS
MINNEAPOLIS

ALL THAT IS
A Naturalistic Faith for the Twenty-First Century

Scripture quotations are from the Revised Standard Version of the Bible, copyright © 1946, 1952, 1971 by the Division of Christian Education of the National Council of the Churches of Christ in the USA. Used by permission. All rights reserved.

Material in Appendixes B and C from Gerd Theissen and Annette Merz, *The Historical Jesus*, and from Arthur Peacocke's *Theology for a Scientific Age*, both published by Fortress Press and SCM Press, are reprinted by permission.

Cover photo: Rila Monastery, Bulgaria © Royalty-Free/Corbis

Library of Congress Cataloging-in-Publication Data
Peacocke, A. R. (Arthur Robert)
  All that is : a naturalistic faith for the twenty-first century / Arthur Peacocke ; a theological proposal with responses from leading thinkers in the religion-science dialogue ; edited by Philip Clayton.
    p. cm. — (Theology and the sciences)
  Includes bibliographical references.
  ISBN-13: 978-0-8006-6227-1 (alk. paper)
  1. Religion and science. 2. Naturalism—Religious aspects—Christianity. I. Clayton, Philip, 1956- II. Title.
  BL240.3.P43 2007
  261.5'5—dc22

2006103474

The paper used in this publication meets the minimum requirements of American National Standard for Information Sciences—Permanence of Paper for Printed Library Materials, ANSI Z329.48-1984.

Manufactured in the U.S.A.

11   10   09   08   07   1   2   3   4   5   6   7   8   9   10

Contents

A Naturalistic Christian Faith for the Twenty-First Century: An Essay in Interpretation

Arthur Peacocke

Responses

Contributors

Donald M. Braxton is the J. Omar Good Professor of Religious Studies at Juniata College. His research focuses primarily on the cognitive mechanisms implicated in religion, the evolution of religion, and religion in an information age. His recent publications have appeared in *Zygon*, the *Metanexus Digest*, and the *Fourth R*.

Philip Clayton is Ingraham Professor at the Claremont School of Theology and Professor of Philosophy and Religion at the Claremont Graduate University. His research focuses primarily on philosophical, metaphysical, and theological issues raised by evolutionary biology and contemporary neuroscience. His recent publications include the *Oxford Handbook of Religion and Science* and *Mind and Emergence* (both by Oxford University Press).

Willem B. Drees is Professor of Philosophy of Religion and Ethics at Leiden University, the Netherlands. He has an advanced degree in theoretical physics and has earned doctorates in theology and in philosophy. He is president of ESSSAT, the European Society for the Study of Science And Theology. Among his publications are *Beyond the Big Bang: Quantum Cosmologies and God; Religion, Science and Naturalism*; and *Creation: From Nothing until Now*; as well as a Dutch translation, with an introduction, of articles by Arthur Peacocke.

Philip Hefner is Professor of Systematic Theology Emeritus at the Lutheran School of Theology at Chicago and editor-in-chief of *Zygon: Journal of Religion and Science*. His most recent works include *Technology and Human Becoming*; *Spiritual Transformation and Healing: Anthropological, Theological, Neuroscientific, and Clinical Perspectives* (edited with Joan Koss-Chioino); and *Religion and Science as Spiritual Quest for Meaning* (in press).

Christopher C. Knight is the Executive Secretary of the International Society for Science and Religion and a Research Associate in the Faculty of Divinity of the University of Cambridge. His recent work has focused on the development of a theological understanding of naturalism, especially in relation to the concepts of revelation and incarnation. His approach to the first of these issues is set out in his book, *Wrestling with the Divine: Religion, Science, and Revelation* (2001), and to the second in *The God of Nature: Incarnation and Contemporary Science* (forthcoming).

Nancey Murphy is Professor of Philosophy at Fuller Theological Seminary, Pasadena, California. Her research interests include the relations between theology and modern and postmodern philosophy; theology and science; and neuroscience and philosophy of mind. Her latest book (with Warren Brown) is *Did My Neurons Make Me Do It? Philosophical and Neurobiological Perspectives on Moral Responsibility and Free Will* (Oxford, 2007).

Ann Pederson is a Professor of Religion at Augustana College and an Associate Adjunct Professor in the Section of Ethics and Humanities at the University of South Dakota Sanford School of Medicine. Her research interests include the connections between constructive theology and music, and between religion and science (particularly medical science). Her recent publications include *God, Creation, and All That Jazz* and *The Music of Creation*, co-authored with Arthur Peacocke.

Karl E. Peters is Professor Emeritus of Philosophy and Religion at Rollins College. He is coeditor of *Zygon: Journal of Religion and Science* and a Past President of the Institute for Religion in an Age of Science. For nearly forty years he has lectured and published on issues in science and religion, with a special interest in understanding how religion and science can be related to everyday living. Many of his reflections are in his book *Dancing with the Sacred: Evolution, Ecology, and God* and his articles in *Zygon*.

Robert John Russell is the Ian G. Barbour Professor of Theology and Science in residence at the Graduate Theological Union, Berkeley, California, and founder and director of the Center for Theology and the Natural Sciences, also in Berkeley. He is the recent editor of *Fifty Years in Science and Religion: Ian G. Barbour and His Legacy* (2004); author of *Cosmology, Evolution, and Resurrection Hope: Theology and Science in Creative Mutual Interaction*, edited by Carl S. Helrich (2006); and General Editor of the series of five research publications on scientific perspectives on divine action sponsored by CTNS and the Vatican Observatory.

Keith Ward is Professor of Divinity at Gresham College, London, and Emeritus Canon of Christ Church, Oxford. He is a Fellow of the British Academy. His main research interests are in comparative theology and religion and science. His recent publications include: *God: A Guide for the Perplexed*, *Pascal's Fire*, and *Re-Thinking Christianity*.

Acknowledgments

This book is dedicated to the memory of Arthur Peacocke. Each of the authors joins with me in expressing a debt of gratitude to the work in science and theology that Arthur accomplished over a rich and productive lifetime. We hold equally deep, though perhaps more diverse, debts to Arthur for the different ways that his life touched and affected us individually as scholars and as persons. While more specific acknowledgments appear in the various chapters, the book as a whole represents a first attempt on our part to honor Arthur Peacocke's life and career.

It was clear to all who knew him that Arthur's work would not have been possible without the support and encouragement of his wife, Rosemary Peacocke. In none of Arthur's writings was her role more significant than in the composition of the "Essay in Interpretation" and "Reflections" that form the core of this volume. I am also grateful to Rosemary for supplying information and feedback that were vital to our preparing this volume for publication, despite the difficulty of these months.

Many other persons have helped to bear the burdens of the editorial process. Philip Hefner offered particular support to this project from the very beginning and played an important role in the contacts with Fortress Press. We thank Kevin Sharpe for agreeing to publish the volume in his Theology and the Sciences series. Michael West, Editor-in-Chief, and the team at Fortress Press are responsible for the appearance of this volume less than two months after submission of the typescript—an unprecedented accomplishment in academic publishing.

A great team of assistants at three different universities provided invaluable research and editorial support on this project. Thanks are due in particular to Simeon Zahl at the University of Cambridge, Clair Linzey at Harvard Divinity School, and Carol Jheri Lee of Santa Rosa, California. Above all I express my gratitude to the Assistant Editor of this volume, Zachary Simpson. This is now the fourth volume that we have produced together during Zach's sojourn at the Claremont Graduate University, and I cannot imagine getting a volume into print without his highly professional assistance.

Finally, I thank the other nine authors, each of whom agreed on extremely short notice to set aside other pressing obligations in order to write a major response to Arthur's Essay in time for him to compose his replies. Many then abridged their original responses, sometimes by upwards of fifty percent, in order to make them the succinct and effective statements that they now are. That people of this stature would set such priority on this task is itself a testimony to the importance and impact of Arthur Peacocke's work.

Philip Clayton
December 2006

Editor's Introduction

Philip Clayton

Theology's relationship to her past is a complex one. On the one hand, theology is an never-ending quest; it calls one continually to respond in novel ways to new "horizons of interpretation," whether they come from science, philosophy, culture, or society. On the other hand, theology harkens continually back to its past—its scriptures, its classic figures, and its traditional themes or *loci*—in an ongoing process of retrieval and renewal.

This highly complex interweaving of past and present is beautifully exemplified in the theology of Arthur Peacocke. In his work Peacocke shows a deep concern with the naturalistic assumptions of the empirical sciences and with the need to find an adequate theological response to them. Indeed, scientific naturalism plays the role not of a "foe" but a "friend"; it calls theologians forward and invites them to develop new ideas that can reduce or eliminate the tensions, ideas such as "emergentist monism" and "panentheism." And yet, while Peacocke's theology embraces naturalism as its current horizon of interpretation, it still seeks to respond to it with resources that have been part of the history of Christian doctrine, practice, and liturgy since the beginning.

For more than a year before his death on October 21, 2006, Arthur Peacocke was working to formulate a final major statement on the relationship between naturalism and Christian faith. As the title indicates, he had concluded that a much more "naturalistic Christian faith" is possible than many theologians are willing to consider. Of course, there are *some* versions of naturalism that, according to Peacocke, would not be compatible with Christian faith, just as there are some understandings of Christianity that are incompatible with even the mildest forms of naturalism. Peacocke's concern always lay not with the extremes but with the overlap set between naturalism and Christian faith. The resulting "Essay in Interpretation," here published for the first time, can thus be read as the attempt to find *a* version of naturalism and *a* Christian theology such that the region of overlap between them is as large as possible. Indeed, the goal of resolving the tensions and apparent conflicts between science and Christian faith would be achieved, he felt, as soon as scholars could agree on an acceptable formulation of the two sides, one that establishes a sufficient degree of overlap between them.

In genuine scientific spirit Peacocke was not satisfied merely to bequeath the world with some final solution. He labeled his work an "essay" in order to emphasize its exploratory and hypothetical character. Though it is in some ways the culmination of his lifelong quest for "paths from science towards God," he also seemed to view the Essay as his first attempt at formulating "a naturalistic Christian faith for the twenty-first century." For this reason, Peacocke wanted his proposal to be placed immediately into the hands of critics, so that they could evaluate what was useful and what was problematic in his arguments; and he insisted that the Essay should only appear in print accompanied by these criticisms. Thus, beginning less than nine months before his death, we began to solicit these responses.

The resulting book is indeed richer than it would have been without these critical interactions. In a way that we could not have anticipated, the ten responses sketch a picture of the broad terrain of *religious naturalism*, and in particular the options for more naturalistic formulations of Christian belief. The result is a remarkably complete spectrum of the possible positions that one can take on this topic.

On the one end of the spectrum are those thinkers who hold that Peacocke has gone too far in one respect or another. (Mindful of the dangers of oversimplifying, I will not attempt to name the various authors in this book who might fall under this and the following categories. It is better that readers identify for themselves which respondents belong in which categories and to what extent.) To cast doubt on the possibility of physical miracles, for example; or to modify classical theism in the direction of panentheism, which fails adequately to specify the ontological distance between God and world; or to deny a dualistic view of humans as unity of body and soul; or to temporalize the being of God rather than conceiving God as outside of time—any of these moves, these critics maintain, will lead to an inadequate formulation of Christian theology.

On the other end of the spectrum are those thinkers who insist that Peacocke has not gone far enough. That he still speaks of a personal God who transcends the physical universe as a whole; that he writes of that God as having specific intentions; that he conceives God as exercising those intentions in the natural world, in evolution, and in human lives; that he seeks to locate ontologically special forms of the presence and influence of God in Jesus Christ, in the church, and in the eucharist, sometimes using the language of incarnation—these are signs, according to the second group of critics, that Peacocke's "naturalization" of Christian faith has not gone far enough.

To make things more complex, note that there is another group of respondents who argue that Peacocke has remained in far greater conformity with classical Christian faith than he realizes. Interestingly, respondents reach this conclusion for two very different sets of reasons. Some argue that the drive to synthesize Christian faith with (something like) naturalism was already present in the early church fathers and has been preserved in the Eastern Orthodox tradition. Thus what Peacocke presents as a modernized form of Christian theism—say, his panentheism—in fact mostly recapitulates what have long been core features of Orthodox theology. Others also argue that Peacocke's position stands in greater continuity with the historic Christian tradition than one might think, but for a different reason: he underestimates how far his position really is from contemporary naturalism. Since he accepts the separate existence of an eternal and transcendent God, speaks of that God's engaging in self-revelation and intentionally exercising influence on the world, is willing to use the language of incarnation, and affirms that humans have a spiritual core that can and does respond to divine influences, his views on human persons, God, and divine providence are significantly more orthodox than he himself may have thought. Subsequent critics will have to evaluate whether Peacocke is less of a revisionist and more orthodox than he thought and, if so, for which reason: because naturalism is more orthodox, or Peacocke less naturalistic, than might at first appear.

A final collection of respondents locate themselves in fundamental agreement with Peacocke's program pretty much as he himself understood it; their criticisms are, as it were, internal to the program. These theologians and philosophers may attempt to nuance the project, to correct it in various ways, or to work out its implications in fuller or bolder fashion. Still, they here endorse *something like* Peacocke's own approach to overcoming the tensions between science and theology. Their challenges are to more specific claims, such as his construal of how God acts on the world as a whole, or how "strong emergence" and "downward causation" work, or how the spiritual resurrection of Jesus is to be understood, or how the eucharist suggests a sacramental view of creation as a whole. One does not find complete agreement among these particular critics: some want to strengthen Peacocke's claims on behalf of mental causation, others to weaken it; some wish to add the possibility of quantum-level divine influence, others to limit the idea of divine influence on the world as a whole; some challenge his modernist assumptions, and others his continuing use of traditional theological terms. Even in the case of such specific challenges to his program, Peacocke seemed to welcome the criticisms, expressing the conviction that they would eventually lead to the formulation of more comprehensive and more adequate theological responses to science.

Whatever one's final evaluation of Peacocke's proposals, one cannot overstate the importance of the topics with which he struggled and on which he wrote over the course of his lifetime. The questions raised here and the theories advanced concern the fundamental features of scientific view of the world as well as the fundamental nature of theology. Whether twentieth-century science gradually underwent a transition from a "reductionist" to an "emergentist" paradigm is a debate of great significance, both historically and in the present—no matter how one finally answers it. Whether theology ought to try to incorporate a naturalistic perspective and, if so, whether it can successfully do this without abandoning beliefs that are indispensable to Christian or Jewish or Muslim religious identity—these are questions that lie at the heart of the struggle to define the nature of theology today. The agreements and disagreements that are played out in these pages between Peacocke's Essay, his ten respondents, and his final "Reflections on the Responses," owe their drama to the high stakes of this debate.

Superlatives are used too lightly, and the tendency is strong to romanticize an individual after his or her death. Nevertheless, it is impossible not to recognize that Arthur Peacocke was one of a very small number of thinkers whose efforts have helped to bring the discussion of religion and science to public attention over the last decades.

Many will write tributes to Arthur Peacocke as a person and will seek to analyze the impact of his thought. Here I wish merely to say a few words about the background to this particular volume and the circumstances under which it came into existence.

A number of people had told me how weak Arthur was at his birthday celebration at the University of Cambridge on July 3, 2005, and how far advanced his cancer was. Thus I was surprised, when visiting him at his home in Oxford in late January 2006, to find him so energetic and so much like himself. He spoke passionately about his recent reflections on the relationship of theology and the natural sciences and about new insights that he had come to. Indeed, it turned out that, somehow, despite battling an illness that had nearly killed him over the preceding months, Arthur had managed to write a lengthy "essay," a major new statement of his position on theology and naturalism. He wanted to find a few scholars who would have the time to respond to this final statement. Might I be willing to help him contact these people, and perhaps to arrange for the publication of the resulting text? Thus began a collaboration that led right up through the last weeks of Arthur life.

It seems appropriate in this context to try to communicate the courage that Arthur showed through the months that followed, the passion to make every bit of progress that he could in the effort to come to more adequate answers, and his commitment to keep thinking and writing as long as there was strength and breath left in his body. Every other week, it seemed, another revision to the Essay arrived, and every other week the name of another possible respondent. Of course, his knowledge that he had very little time left added an urgency to this drive to complete the project; and yet, in another respect, Arthur was just doing what he had always done.

By the time that the last of the ten responses was in, Arthur was already quite weak. He had faced multiple hospitalizations, and the pain was continuous and often debilitating. He emailed at one point, with typical understatement, "the pain has been assuaged a little by the last treatment—or so I hopefully guess—but I have good and bad times, needless to say." Later he wrote, with a touch of humor, "I am experiencing reduced energy for doing anything—let alone responding to critical friends!" As a result, Arthur warned me that he would only be able to write a short, general reply to the ten responses, since he no longer had the strength left to address them all. I encouraged him to respond to each one in whatever way he could, and promised that we would assemble his various replies and include them in the book.

Somehow—finding inner resources in a way that I still do not understand—he did exactly that. Each week, as his condition rapidly deteriorated, another few pages of text would arrive. By the time that he went into the hospital in July, all the replies were written. And even then he was not satisfied. Revisions to the replies kept coming. Even in the last weeks, when Arthur was extremely weak, I would receive emails, written for example by his son-in-law, reporting on the most recent conversation with Arthur and transmitting his request that a certain phrase be added or changed. Literally as long as energy remained in his body and as long as he was able to talk, Arthur was working on refining and elaborating his position, leaving a steadily more detailed argument to which other scholars might later respond.

Yet all of this occurred with an amazing lack of ego. His wife, Rosemary, wrote after his death that Arthur would have been "very surprised at the honour you and others are according him," and saying that he would not have expected any such response. I believe that what she wrote is true. Rather than viewing himself as bequeathing to the world a final solution to the issues, Arthur Peacocke saw himself as merely another participant, his limitations all too evident to himself, in an ongoing dialogue about some of the most urgent questions that humanity faces in an age of science. One cannot help but think of the Reformation phrase, *semper reformata* ("always reformed, always reforming") in connection with his attitude.

Clearly the "Essay in Interpretation" that forms the centerpiece of this book does not provide the last word on theology and naturalism. Still, it is a bold and powerful proposal. It asks scientists as well as theists to accept revolutionary changes in their self-understanding, knowing that a genuine rapprochement between science and theology will be possible only if both sides are prepared to do so. The proposal is courageous, because it will evoke criticism from both sides; it also shows tenacity, because it continues to chip away at themes with which Peacocke wrestled since the beginning of his career. Above all, however, it is deeply dialogical. At the author's own request the Essay appears side by side with a series of criticisms by some of the leading figures in the field, in a book in which the critics' voices constitute a larger portion of the text than the author's own.

A few weeks before his death, Arthur composed and circulated a final statement. He entitled it "Nunc Dimittis," the Latin phrase that opens Simeon's canticle in Luke 2: 29-35: "Lord, now lettest thou thy servant depart in peace, according to thy word: For mine eyes have seen thy salvation, which thou hast prepared before the face of all people." Fortress Press has agreed to exclude the "Nunc Dimittis" from the copyright that governs the rest of the book, so that it can be reproduced and circulated by those who wish to do so. It is appropriate to close with Arthur's words, which represent his final understanding of "the end of all our exploring":

> This [final illness] is a new challenge to the integrity of my past thinking. I am only enabled to meet this challenge by my root conviction that God is Love as revealed supremely in the life, death and resurrection of Jesus the Christ. However the fact remains that death for me is imminent and of this I have no fear because of that belief. . . . I know that God is waiting for me to be enfolded in love.

A NATURALISTIC CHRISTIAN FAITH FOR THE TWENTY-FIRST CENTURY: AN ESSAY IN INTERPRETATION

ARTHUR PEACOCKE

Preface

Reflection on one's past writings often allows one to see them in a new perspective. The following "essay in interpretation" is the result of my perceiving a certain connectedness between the theology I have been developing,[1] construed in relation to the scientific account of the world, and the way the natural world itself, as perceived through the sciences, may be interpreted.[2] The hierarchy of complexity of the natural world, increasingly explicated by the sciences both in detail and through wider concepts, has made apparent how new realities emerge at higher levels of complexity, with all their interactions and ramifications and how these higher levels of complexity can influence, and even transform, the behavior of the lower-level entities that constitute them. It has occurred to me that this same "scenario," if I may so put it, is also manifest in those situations we denote as spiritual or religious experiences, which theology then attempts to analyze and to formulate intellectually and conceptually.

In this extended essay I have tried to work this out in more detail. The attempt has necessarily led me to repeat some earlier theological formulations of mine, and I have (shamelessly) self-plagiarized revised parts of earlier texts, scattered over the years in various books and papers, to bring the relevant ideas together again in a new context. Such a re-assembling of earlier ways of expressing my thoughts has, I have found, a certain value, in that it clarifies in a new context the perspective one has implicitly been adopting along the way. It also allows these earlier formulations of my theology to be given a more comprehensive and, I would hope, a more comprehensible form.

In engaging in this exercise, it has become clearer to me that, all along and presumably because of my *Bildung* as a scientist, I have been seeking as *naturalistic*[3] a formulation of the content of Christian faith as is achievable, while still doing due justice to the data on which that theology rests. Others have

3

also and more explicitly had such an aim, of course. I find that I have myself been adopting a form of naturalism,[4] insofar as I stress the completeness and reality of the natural world of the sciences. At the same time, and unlike some of these other authors, I have allowed the term "God" to be referential—that is, to affirm that there is an Ultimate Reality named as "God" and that "God" does not merely denote a feature of our experience interpreted existentially.[5] The pressure to retell Christian faith in some kind of naturalistic terms is, not surprisingly, endemic in our present cultural situation (as I have outlined in the Prologue), for the scientific account of the natural world, including human beings, is the best-established account of the realities in which we are embedded.

I cannot pretend that this work is a systematically complete whole, as is the aim of most "systematic theologians," but I would hope that its open-endedness, its porosity at its boundaries and rough edges, is, in practice, more in tune with the spiritual explorations of many in our times than are the cut-and-dried offerings of many systematic theologies. So I invite the reader to join me in this essay in interpretation, hoping that it will often open new vistas and insights for them, as it has for me.

Chapter 1

Prologue:
Naturalism, Theism, and Religion

A host of surveys indicates that what Christians, and indeed other religious believers, today affirm as real fails to generate any conviction among many of those in Europe and in intellectual circles in North America, who seek spiritual insight and who continue regretfully as wistful agnostics in relation to the formulations of traditional religions. Many factors contribute to this state of affairs. But one of these, I would suggest, is that the traditional language in which much Christian theology, certainly in its Western form, has been and is cast is so saturated with terms which have a supernatural reference and color that a culture accustomed to think in naturalistic terms, conditioned by the power and prestige of the natural sciences, finds it increasingly difficult to attribute any plausibility to it. Be that as it may, there is clearly a pressing need to describe the realities that Christian belief wishes to articulate in terms that can make sense to that culture while expanding its theological significance.

Correspondingly, there is also a perennial pressure, even among those not given to any form of traditional religiosity, to integrate the understandings of the natural world afforded by the sciences with the very real, widespread "spiritual" experiences that include both interactions with other persons and awareness of the transcendent.

Both of these pressures in contemporary life accentuate the need to find ways of integrating "talk about God," that is, theology, and those experiences denoted as "spiritual" or "religious" with the worldview engendered and warranted by the natural sciences. For the "god of the gaps" and the whole notion of divine "revelation" have become questionable in such a worldview. Yet, I will argue, this move is not equivalent to the expulsion of the creator God of the monotheistic religions. Is it possible to recognize the cogency and comprehensiveness of the scientific worldview while also believing in a creator God—in particular, is it possible to formulate a Christian *naturalistic* theology? It is not so much a matter of conjecturing the possibility of such a formulation—there is, rather, a cogent imperative to engage in such a project. For Christian theology (and indeed that of all the Abrahamic faiths) places a strong emphasis on the doctrine of creation, making the claim that "nature" is the theological locus of God's grace. Thus, the first Christian affirmation about the Godhead in the ancient (so-called) Apostles Creed is that God is "the Maker of heaven and earth," amplified in the Nicene Creed to refer to "all things visible and invisible." On this premise, one would expect the created world to reflect in its very nature the purposes of God, its Creator, and how God and God's relation to the created world are best to be articulated. Only when this foundation of insights into the nature of God and God's relation to the world has been laid should it be possible to develop an understanding of the significance of the historical Jesus of Nazareth—that is, an account of Jesus as the Christ of faith.

Too little of such exposition has taken account of the rapidly growing understanding of the natural world afforded by the contemporary sciences, if only in its broadest implications. These include *inter alia* the ubiquity of regularity in the world, its closure to non-natural forces, the scepticism this induces of the supernatural, and the emergentist character of natural, creative processes. Too often, those attempting to develop a Christian theology that takes account of the worldview engendered by the sciences have been content to leave intact relatively traditional formulations of the Christian faith. Indeed, they have often been concerned explicitly to defend and entrench these formulations, while not really recognizing at any deep level the need for the radical revisions that are necessary if coherence and intellectual integrity are to be achieved.

One of the most cogently and thoroughly argued attempts to formulate a Christian naturalistic theology is that of Charles Hardwick,[6] which is based on his conviction that, "Broadly construed, supernaturalism has been the basis of most Western religious views. It asserts that the world of nature fails to exhaust the 'real' because reality consists of nature and a superordinate reality that grounds the natural world and provides its end."[7] He contrasts naturalism with supernaturalism by listing a number of "family resemblances"[8] among various forms of naturalism, each of which diverges from a parallel element in supernaturalism. These family resemblances are:

(1) that *only* the world of nature is real; (2) that nature is necessary in the sense of requiring no sufficient reason beyond itself to account either for its origin or ontological ground; (3) that nature as a whole may be understood without appeal to any kind of intelligence or purposive agency; and (4) that all causes are natural causes so that every natural event is itself a product of other natural events.[9]

By "nature" I surmise he means simply all-that-is, as described and explicated by the sciences, including humanity and society—no supernaturalism, no entities not accessible to critical and experimental inquiry. (1) and (4) are the general presuppositions of many "naturalisms," with (1) precluding the attribution of any reference to the word "God"—there is no such reality, it asserts. Both (2) and (3) are, in fact, metaphysical assumptions made by Hardwick and are therefore not deducible from the content of the sciences as such: (2) is disputable, in its ontological form being the question of the existence of a creator God, and (3) is clearly not demonstrable. He, and others with a more anti-theological stance, clearly challenge whether any form of naturalism can be coherent with a belief in an independently existing creator God.

Hardwick also argues that naturalism, as defined by (1) to (4) above, poses as much of a challenge to many contemporary revisionary theisms as it does to classical theism, all of which, he claims, share the following broad commonalities: (a) that God is personal, (b) that some form of cosmic teleology is metaphysically true, and (c) that there is a cosmically comprehensible conservation of value.[10] However, Ursula Goodenough[11] can call herself a "religious naturalist," for she assumes that religion can be regarded only as that which "provides the means for personal wholeness and social coherence."[12] For a religious naturalist such as her, meaning is inherent in nature, and the search for it does not require an independently existing God.

Karl Peters has, on the other hand, developed a "religious naturalism" that does not shy away from the use of the word "God," by which he means not a being who creates the world but the process of creation itself.[13] For him the sacred (or God) is the "creative activity of nature, human history, and individual life."[14] His "naturalism means that everything is energy-matter and the information according to which energy-matter is organized. . . . [T]he causes of things are not personal, mental and intentional—except when personal creatures such as humans and probably some animals are involved."[15] In spite of this, he claims that his use of the term "God" has an ontological reference to be found, not indeed in any personal entity, but in his notion of creativity as rooted in the combination of the Darwinian process (of random variation with natural selection) and non-linear, non-equilibrium thermodynamics, which supports the natural, deep-seated process of the creation of complexity in the natural world. This is indeed a *naturalistic* view concerning the character of the "sacred," which Peters uses as almost synonymous with "God." He asks, "Can one observe God?" and replies, "I think so, if one considers the cues[16] themselves as part of the creative process. . . Then we don't need to say 'Dance with me' or 'Follow my lead, my music of the spheres.' The invitation needs only to be 'Come dance.' God *is* the music. Responding only to this brings one into relation with our sacred center."[17]

David R. Griffin takes a different stance from both Hardwick and Peters, one that is akin to that of Goodenough, though subtly different in being more explicitly theist. He too defines naturalism in what he calls a "minimal and generic sense" as:

> the doctrine that the universe involves an extremely complex web of cause-and-effect relations; that every event occurs within this web, having causal antecedents and causal consequences; and that every event exemplifies a common set of causal principles. . . . Naturalism in this minimal sense does not entail atheism. There are theistic versions of naturalism. Naturalism in this minimal sense rules out only supernaturalism, defined as the doctrine that there is a supernatural being existing outside the otherwise universal web of cause-effect relations, which can violate it. . . I call naturalism in this generic or minimal sense "naturalism$_{ns}$" with the "ns" standing for "nonsupernaturalistic."[18]

Griffin further distinguishes this sense of "naturalism," to which I also adhere in this work, from the doctrine (basically [1] and [4] above) that "nature is all there is," with "nature" understood to mean the totality of all finite things, processes, and events. Using "nati" to stand for "nature is all there is," he refers to this doctrine as "naturalism$_{nati}$. . . . Naturalism in this sense, by specifying that there is nothing other than the totality of finite things, does rule out theism."[19] Griffin has surveyed historically[20] how "the great truth" of scientific naturalism has emerged in its present form of "naturalism$_{sam}$," with "sam" standing for "sensationalist-atheistic-materialistic."

> By becoming embodied in a great falsehood, naturalism$_{sam}$, this great truth [of naturalism$_{ns}$] has appeared in our time in a grossly distorted form. . . . Scientific naturalism in the generic sense can be regarded as a great truth of universal importance. . . . In so far as theologians reject scientific naturalism totally—without distinguishing naturalism$_{ns}$ from naturalism$_{sam}$—they are rejecting a great truth as well as creating an insuperable conflict between theology and science.[21]

With this I concur.

This essay is based on a theology, which I have developed elsewhere[22] as consistent with scientific perspectives and with the view of the world implicit in natural science. This theology may be properly deemed to be "naturalistic" insofar as it assumes that the world of nature is real, that science unveils its realities, and that this natural world contains those entities, processes, and structures that are explicable and eventually rendered intelligible through the natural sciences—hence, consistent with Hardwick's naturalistic proposition

(4), no "supernatural" entities, no "miracles" that break the laws or regularities of nature discovered by science, no dualisms within the natural world. Indeed, I also follow Hardwick's strategy, which rests on his (and my) "conviction that theologians must take far more seriously the possibility that naturalism provides the true account of our world, and indeed in its materialist or physicalist version."[23] I agree with him that "theologians too readily—and too facilely—dismiss philosophical naturalism" and with his dismissal of their assumption that it is philosophically deficient. Likewise, I share his aim of showing that its truth would not "render the Christian witness of faith impossible"—provided that that faith does not continue to be served up as "tired eschatological symbols . . . largely unreflectively in mythological forms that scarcely any educated person believes today."

However the theology espoused here also affirms that the natural, including the human, world owes its existence to another entity, a Creator God, who is real (contrast Hardwick's [1]), "at least personal" (see Appendix A) and has purposes for this world (contrast Hardwick's [3]). Yet it maintains that God does not implement these purposes through "miracles" that intervene in or abrogate the world's natural regularities, which continue to be explicated and investigated by the natural sciences. The only dualism which such a stance accepts is indeed that between God and all-that-is, the "world"; it rejects any dualisms within the natural world itself, including humanity. The question will therefore arise of how God could, if God so intended, influence particular events in the world without contravening the regularities established by science (a question addressed in Chapter 9). There is a rejection, in this theology, of any notion of the "supernatural" as a kind of go-between between God and the world, including humanity. I am encouraged to believe such a stance might transpire to be not entirely uncongenial to Christian thought by the fact that the "Eastern tradition knows no such supernatural order between God and the created world, adding, as it were, to the latter a new creation. It recognizes no distinction, or rather division, save that between the created and the uncreated. For the Eastern tradition the created supernatural has no existence."[24]

This theology is also based on an evolutionary perspective of the cosmological and biological sciences. This view entails an understanding of creation by God as a continuous activity, so that dynamic models and metaphors of divine creation and creativity become necessary. The work of God as Creator is regarded as manifest all the time in those very natural processes that are unveiled by the sciences in all their regularities.

Such cold and abstract statements cannot do justice to the deep aesthetic and spiritual significance of this apprehension of the natural world that I intuit, so I resort to the golden prose of a seventeenth-century divine, Thomas Traherne, which expresses, for me, the inner spirit of this perception:

The WORLD is not this little Cottage of Heaven and Earth. Though this be fair, it is too small a Gift. When God made the World He made the Heavens, and the Heavens of Heavens, and the Angels, and the Celestial Powers, These also are parts of the World: So are all those infinite and eternal Treasures that are to abide for ever, after the Day of Judgment. Neither are these, some here, and some there, but all everywhere, and at once to be enjoyed. The WORLD is unknown, till the Value and Glory of it is seen: till the Beauty and the Serviceableness of its parts is considered. When you enter into it, it is an illimited field of Variety and Beauty: where you may lose yourself in the multitude of Wonder and Delights. But it is an happy loss to lose oneself in admiration at one's own Felicity: and to find GOD in exchange for oneself. Which we then do when we see Him in His Gifts, and adore His Glory.[25]

This is, in fact, a spirit deeply rooted in the Christian experience—from the speech attributed to St. Paul on the Areopagus when, quoting a Greek poet, he is said to have declared to his Athenian audience his allegiance to "The Unknown God" of whom it may be said "in him we live, and move, and have our being,"[26] to Dante's celebration of "His glory, in whose being all things move,/ pervades Creation and, here more there less/ resplendent, shines in every part thereof."[27] And not only the Christian experience, for in an even more ancient text, the *Bhagavadgita*, the Divine affirms, "He who sees Me everywhere and sees all in Me; I am not lost to him nor is he lost to Me."[28] As S. Rhadhakrishnan comments on this verse:

It is personal mysticism as distinct from the impersonal that is stressed in these tender and impressive words: "I am not lost to him nor is he lost to Me." The verse reveals the experience of the profound unity of all things in One who is the personal God. The more unique, the more universal. The deeper the self, the wider is its comprehension. When we are one with the Divine in us, we become one with the whole stream of life.[29]

In what follows, an attempt will be made to develop features of a more specifically *Christian* faith that are consistent with such a naturalistic perspective, different from that of Hardwick, but still, I would argue, meriting the designation of "naturalistic." At the same time, the outcome will diverge from that of Hardwick's naturalistic theology with respect to his existentialist interpretation of "God" and "God exists," an interpretation that precludes any real ontological reference to what theology basically affirms. Thus, I attribute ontological status to theological terms through the warrant of a critical realism, in both science and theology, which I have often defended.[30]

While theology has been attempting to adjust itself to the challenges from the scientific worldview, that worldview has itself not been static; in particular, it has come to discern *inter alia* the ability of natural complexes to manifest new properties not associated with their components. Such complexes display, for example, emergent, self-organizing properties. Awareness of this fact, and of the general role of the transfer of information in complex systems, has generated a metaphysics that takes account of this feature of the natural world, namely, *emergentist monism*. Our whole understanding of "the nature of Nature," if one may so put it, is being transformed as we realize that emergent properties arise as a consequence of relationships in complex systems. Even if such systems can be reduced, in one sense, to their components (basically, and monistically, what the physicists tell us underlies matter / energy / space / time), still there is almost always "something more" to be said than the mere identification of those components. When one realizes that all such systems are themselves incorporated into larger and more comprehensive complexes, so that the world (physical, biological, ecological, human, social, cultural) is a System-of-systems and that this is the basis of the fecundity and diversity of the natural world, then any theistic affirmation—any "doctrine of creation"—must include an overt recognition that this is the mode of the divine creativity in nature. For the idea of emergence is becoming central to our understanding of the dynamics and evolution of the natural world, including human beings. Natural systems, it transpires, have an inbuilt capacity to produce new realities; hence any theistic understanding has to recognize that this is the mode and milieu of God's creative activity. Such a basically naturalistic perspective is here linked with a critical realism that takes theological terms to refer, even if putatively, to realities and thus differs from Hardwick's scepticism in this regard. Hence, in the following, "God" will be taken to have a realistic reference, such that it will be appropriate to investigate the relationships and any emergence in the complex of {God + nature + humanity}.

There has also been an impetus under pressure from the scientific worldview to revise ideas concerning God's relation to the world in terms both of a *theistic naturalism* and of *panentheism*. The aim of this essay is to integrate these various developments and to show that this process allows one to formulate theological language in such a way as to escape the cultural impasse described above. This serves to support the claim of theology to refer realistically to the interrelationships of God, nature, and humanity, which are the content of spiritual and religious experience and which are the focus of theology's critical reflection on it. Before we proceed on this quest it is necessary to recall, in Part One, the content and origin of these three naturalistic stances in philosophy and theology that have had to be adopted under this very real pressure from established scientific perspectives.

Chapter 2

Emergentist Monism

The natural and human sciences increasingly give us a picture of the world[31] as consisting of a complex series of levels of organization and matter in which each successive member of the series is a whole constituted of parts preceding it in the series.[32] The wholes are organized systems of *parts* that are dynamically and spatially interrelated, a feature sometimes called a "mereological" relation. Furthermore, the properties distinctive of the "wholes"—the organized complexes of the "parts"—result from the particular forms of interrelatedness of those parts which, in isolation as microphysical entities, also have their own particular properties. This feature of the world is now widely recognized to be of significance in relating our knowledge to its various levels of complexity—that is, the sciences which correspond to these levels.[33] It also corresponds not only to the world in its present condition but also to the way complex systems have evolved over time out of earlier, simpler ones. Hence this sequence of new relationships, of wholes/parts, can be observed when moving, synchronically, up the ladder of complexity as it is now or, diachronically, through cosmic and biological evolutionary history.

I shall presume at least this with the "physicalists": all concrete particulars in the world (including human beings)—with all of their properties—are constituted only of fundamental physical entities of matter/energy at the lowest level and manifested in many layers of complexity. One might call it a "layered" physicalism. This is indeed a *monistic* view (one that is reductionist in the constitutive-ontological sense), in the sense that everything can be broken down into whatever physicists deem ultimately to constitute matter/energy (e.g., quarks, superstrings?). No extra *entities* or *forces,* other than the basic four forces of physics, are to be inserted at higher levels of complexity in order to account for their properties. However, what is significant about natural processes and about the relation of complex systems to their constituents is that the concepts needed to describe and understand—as indeed also the methods needed to investigate each level in that hierarchy of complexity—are specific to and distinctive of those levels. It is very often the case (but not always) that the properties, concepts, and explanations used to describe the higher-level wholes are not logically reducible to those used to describe their constituent parts, themselves often also constituted of yet smaller entities. This is an epistemological assertion of a non-reductionist kind.

When the epistemological non-reducibility of properties, concepts, and explanations applicable to higher levels of complexity is well-established, their employment in scientific discourse can often, *but not in all cases*, lead to a putative and then to an increasingly confident attribution of reality to that to which the higher-level terms refer. "Reality" is not confined to the physico-chemical alone. One must accept a certain "robustness" (W. C. Wimsatt[34]) of the entities postulated or, rather, discovered at different levels, resisting any attempts to regard them as less real in comparison with some favoured lower level of "reality." Each level has to be regarded as a cut through the totality of reality, if you like, in the sense that we have to take account of its mode of operation at that level. New and distinctive kinds of realities at the higher levels of complexity may properly be said to have *emerged. What* emerges involves relationships between the wholes and the parts which cannot be explicated by physics (including its application at the micro-physical level) or, indeed, by any of the sciences applicable to levels lower than the particular one under consideration.

Much of the contemporary discussion of reductionism has concentrated upon the relation between already established theories pertinent to different levels. This way of examining the question of reductionism is less appropriate when the context is that of the biological and social sciences, for which knowledge hardly ever resides in theories with distinctive "laws." In these sciences what is more often sought is usually a *model* of a complex system that explicates how its components interact to produce the properties and behavior of the whole system—organelle, cell, multi-cellular organism, ecosystem, etc. These models are not presented as sentences involving terms that must be translated into lower-level terms for reduction to be successful; rather, they are visual systems, structures, or maps representing multiple interactions and connecting pathways of causality and determinative influences between entities and processes. When the systems are not simply aggregates of similar units, then it can turn out that the behavior of the system is due principally, sometimes entirely, to the distinctive way its parts are put together—which is what models attempt to make clear. This incorporation into a system constrains the behavior of the parts and can lead to behavior of the system as a whole that is often unexpected and unpredicted. As W. Bechtel and R. C. Richardson[35] have expressed it: "They are *emergent* in that we did not anticipate the properties exhibited by the whole system given what we knew of the parts" (267). They illustrate this from a historical examination of the controversies over yeast fermentation of glucose and oxidative phosphorylation. What is crucial here is not so much the unpredictability but the inadequacy of explanation if only the parts are focused upon, rather than the whole system. "With emergent phenomena, it is the interactive organization, rather than the component behavior, that is the critical explanatory feature" (285).

There are, therefore, good grounds for utilizing the concept of emergence in our interpretation of naturally occurring, hierarchical, complex systems constituted of parts which themselves are, at the lowest level, made up of the basic units of the physical world. I shall denote[36] this position as that of *emergentist monism,* but in doing so I am not espousing those earlier forms of monism according to which reality itself is all of one *kind.* The position here is intended to recognize that, as noted above, everything in nature can indeed be broken down into whatever physicists deem to constitute matter/energy. It could be called "emergentist *materialism*" but for the fact that matter is interchangeable with energy at the deepest level and that the fundamental building blocks of the natural world may have to be described in quite other terms, e.g., as superstrings. The term "monism" is also emphatically *not* here intended (as is apparent from the non-reductive approach that I adopt) in the sense in which it is taken to mean that physics, starting from the micro-physical level, will eventually explain everything (which is what, for example, "physicalism" is usually taken to mean). *Emergentist monism* affirms that natural realities, although basically physical, evidence various levels of complexity with distinctive internal inter-relationships between their components, such that new properties, and also new realities, emerge in those complexes—in biology in an evolutionary sequence.

If we do make such an ontological commitment about the reality of the "emergent" whole of a given total system, the question then arises how one is to explicate the relation between the state of the whole and the behavior of parts of that system at the micro-level. A number of related concepts have, in fact, been developed in recent years to describe the relations between the higher levels of complexity and the lower ones in both synchronic and diachronic systems—that is, both those in some kind of steady state with stable characteristic emergent features of the whole and those that display an emergence of new features over the course of time.

It transpires that extending and enriching the notion of causality now becomes necessary because of new insights into the way complex systems in general, and biological ones in particular, behave. For in recent years we have gained a subtler understanding of how higher levels can be a factor in what is happening at the lower levels. These developments allow the application in this context of the notion of a determining ("causal") relation from whole to part (of system to constituent)—never ignoring, of course, the "bottom-up" effects of parts on the wholes. Such systems depend on the properties of the whole for the parts behaving as they do; the parts are what they are in the new, holistic, complex, interacting configurations of that whole. For example, the term *downward causation* or *top-down causation* was employed by Donald Campbell[37] to denote the way in which the network of an organism's relationships to its environment and its behavior patterns over the course of time together determine the actual DNA sequences that are present in an evolved organism at the molecular level —even though, from the "bottom-up" viewpoint, a molecular biologist will tend to describe the form and behavior of that organism once in existence as a consequence of those same DNA sequences. One should perhaps better speak of *determinative influences* rather than of "causation," since the latter can have misleading connotations.

Other systems could be cited,[38] such as the Bénard phenomenon and certain auto-catalytic reaction systems (e.g., the famous Zhabotinsky reaction and glycolysis in yeast extracts), which spontaneously display, and often after a time interval from the point when first mixed, rhythmic temporal and spatial patterns the forms of which can even depend on the size of the containing vessel. Harold Morowitz[39] has indeed, somewhat boldly, identified some twenty-eight emergent levels in the natural world. Many examples of dissipative systems are now also known which, because they are open, far from equilibrium, and non-linear in certain essential relationships between fluxes and forces, can display large-scale patterns in spite of random motions of the units—"order out of chaos," as Prigogine and Stengers dubbed it.[40]

In these examples the ordinary physico-chemical account of the interactions at the micro-level of description simply cannot account for the observed phenomena. It is clear that what the parts (in the Bénard and Zhabotinsky cases, molecules and ions) are doing, and the patterns they form are what they are *because* of their incorporation into the system-as-a-whole; in fact, these are patterns *within* the systems in question. The parts would not be behaving as observed if they were not parts of that particular system (the "whole"). The state of the system-as-a-whole is influencing (i.e., acting like a "cause" on) what the parts, the constituents, actually do. Many other examples of this kind could be taken from the literature, not only involving self-organising and dissipative systems but also, for example, economic and social ones. Terrence Deacon has usefully categorized the different kinds of emergent levels involved in these phenomena.[41]

A wider use of "causality" and "causation" than Humean temporal, linear chains of causality as previously conceived ($A \rightarrow B \rightarrow C \ldots$) is therefore now needed to include the kind of whole-part, higher- to lower-level relationships that the sciences have themselves recently been discovering in complex systems, especially the biological and neurological ones. Such systems manifest a network of inter-relationships between constituent units that are character-istic only of the complex as a whole. I will generally use the term *whole-part influence* to represent the net effect of all those ways in which a system-as-a-whole, operating from its "higher" level, is a determining factor in what happens to its parts, the "lower" level.[42] A holistic state, in this understanding, is determined by (is "caused by" or is "a consequence of") a preceding holis-tic state *jointly* with the effects of its constituents, including their individual properties in isolation; it is therefore not only the lower levels that are doing the real work. Be it noted that such a "joint" effect may *sometimes* be inter-preted as a transmission of "information,"[43] when information is conceived in its broadest sense as that which influences patterns and forms of the organi-zation of constituents. These ways in which inter-relationships in complex systems in the natural world have been explicated provide clues, it will later be urged, to concepts helpful in relating the constituent entities of the God-nature-humanity relation.

This analysis of the determinative ("causal") relationships between wholes and parts has, furthermore, provided a substantial ground for attributing reality to higher-level properties and the organized entities associated with them. We noted that the possession of a distinctive determinative ("causal") efficacy on the part of the complex wholes has the effect of making the sepa-rated, constituent parts behave in ways they would not do if they were not part of that particular complex system (that is, in the absence of the inter-actions that constitute that system). Where such determinative influences of the whole of a system on its parts occurs, one is justified in attributing real-ity to those emergent properties and features of the whole system that have those consequences. Real entities have influence and play irreducible roles in adequate explanations of the world; this influence and role might well include their effects on the environment of the system in question, which should thus also be investigated. *To be real is to have causal power,* it has been reasonably claimed.[44] Such "causal powers" may take time to become manifest, and their absence may not always denote non-reality, so this may be only a sufficient rather than a necessary criterion of the "real." Such powers may not be imme-diately demonstrable, but, when they are, new causal powers and properties can then properly be said to have *emerged.*

Chapter 3

Theistic Naturalism

The incessant pressure from the popular, widespread acceptance of the cogency of science has been toward assuming, with little further consideration, that "the world can best be accounted for by means of the categories of natural science (including biology and psychology) without recourse to the super-natural or transcendent as a means of explanation," which is a definition of "naturalism."[45] According to the dictionary, naturalism is "a view of the world, and of man's relation to it, in which only the operation of natural (as opposed to supernatural or spiritual) laws and forces is assumed."[46] Such a stance seems implicitly to preclude the divine. This inference is not a necessary consequence of taking account of the scientific knowledge of the world and its development, however, for a *theistic* naturalism may be expounded according to which natural processes, characterized by the laws and regularities discovered by the natural sciences, are themselves actions of God, who continuously gives them existence.

Contrary emphases have long historical roots since, for a century or more after Newton, there was still a tendency to think of creation as an act at a specific point in time. At that time, and within the framework of an already existing space, it was believed, God created something external to God's self—not unlike the famous Michelangelo depiction of the creation of Adam on the ceiling of the Sistine Chapel. This led to a conception of God which was very "deistic": God as external to nature, dwelling in an entirely different kind of "space" and being of an entirely different "substance," which by definition could not overlap or mix with that of the created order. In practice, and in spite of earlier theological insights, there was an excessive emphasis on God's transcendence and on the separation of God from what is created. This led to the dominance of *deism*, "the belief in the existence of a supreme being who is regarded as the ultimate source of reality and ground of value but as not intervening in natural and historical processes by way of particular providences, revelations and salvific acts."[47]

Cracks in this conceptual edifice began to appear in the late eighteenth century and early nineteenth century, however, when the age of the Earth inferred from geological studies began to be stretched from the traditional date of 4004 B.C.E., deduced by adding up the ages of the biblical patriarchs. Now "creation" transpired to be a process lasting many hundreds of thousands of years or much more. But it was Charles Darwin's eventually accepted proposal of a plausible mechanism for the changes in living organisms that led to the ultimate demise of the external, deistic notion of God's creative actions and to a renewed emphasis on God's immanence in the created natural world. In particular, those Anglican theologians who were recovering a sense of the sacramental character of the world stressed God's omnipresent creative activity in that world. Thus Aubrey Moore wrote in 1889:

> The one absolutely impossible conception of God, in the present day, is that which represents him as an occasional visitor. Science has pushed the deist's God further and further away, and at the moment when it seemed as if he would be thrust out all together Darwinism appeared, and, under the disguise of a foe, did the work of a friend. It has conferred upon philosophy and religion an inestimable benefit, by showing us that we must choose between two alternatives. Either God is everywhere present in nature, or He is nowhere.[48]

He and his co-religionists were not alone—the evangelical Presbyterian Henry Drummond saw God as working all the time through evolution:

> Those who yield to the temptation to reserve a point here and there for special divine interposition are apt to forget that this virtually excludes God from the rest of the process. If God appears periodically, He disappears periodically. . . . Positively, the idea of an immanent God, which is the God of Evolution, is definitely grander than the occasional Wonderworker, who is the God of an old theology.[49]

Similarly, the Anglican Evangelical Charles Kingsley did not depart far, if at all, from the position of Frederick Temple, the Archbishop of Canterbury, when he affirmed in *The Water Babies* that "God makes things make themselves."[50]

For a theist, God must now be seen as acting creatively in the world often through what we call "chance" or random processes, thereby operating *within* the created order, each stage of which constitutes the launching pad for the next. The Creator unfolds the created potentialities of the universe in and through a process in which its possibilities and propensities become actualized. God may be said to have "gifted" the universe, and goes on doing so, with a "formational economy," which is the set of all of the dynamic capabilities of matter and material, physical and biotic systems that are "sufficiently robust to make possible the actualization of all inanimate structures and all life forms that have ever appeared in the course of time."[51]

Consequently, a revived emphasis on the *immanence* of God as Creator "in, with and under" the natural processes of the world unveiled by the sciences becomes imperative if theology is to be brought into accord with all that the sciences have revealed since those debates of the nineteenth century. A notable aspect of the scientific account of the natural world in general is the seamless character of the web that has been spun on the loom of time; at no point do modern natural scientists have to invoke any non-natural causes to explain their observations and inferences about the past. The processes that have occurred can be characterized as displaying *emergence*: new forms of matter, and new organizations of these forms themselves, appear over the course of time, and these have new properties and new whole-part, determinative influences. It is thus appropriate to affirm that new kinds of reality have emerged in time.

Hence the scientific perspective of the world, especially the living world, inexorably impresses upon us a *dynamic* picture of the world of entities, structures, and processes involved in continuous and incessant change and in process without ceasing. This has impelled us to re-introduce a dynamic element into our understanding of God's creative relation to the world. This was always implicit in the Hebrew conception of a "living God," dynamic in action, but it had been obscured by the tendency to think of "creation" as an event in the past. God has again to be conceived as continuously creating, continuously giving existence to, what is new. God is creating at every moment of the world's existence in and through the perpetually-endowed creativity of the very stuff of the world.

All of this reinforces the need to reaffirm more strongly than at any other time in the Christian (and Jewish and Islamic) traditions that in a very strong sense God is the *immanent* Creator creating in and through the processes of the natural order. The processes are not themselves God but are the *action* of God-as-Creator—rather in the way that the processes and actions of our bodies as psychosomatic persons express ourselves. God gives existence in divinely-created time to a process that itself brings forth the new, and God is thereby creat*ing*. This means we do not have to look for any *extra* supposed gaps in which, or mechanisms whereby, God might be supposed to be acting as Creator in the living world.

A musical analogy may help. When we are listening to a musical work, say, a Beethoven piano sonata, there are times when we are so deeply absorbed in it that for the moment we are thinking Beethoven's musical thoughts with him. If, however, anyone were to ask at that moment (unseemingly interrupting our concentration!), "Where is Beethoven now?" we would have to reply that Beethoven-as-composer was to be found only in the music itself. Beethoven-as-composer is/was other than the music (he "transcends" it), but his interaction with and communication to us is entirely subsumed in and represented by the music itself—he is immanent in it and we need not, and cannot, look elsewhere to meet him in that creative role.[52] Similar interpretations also apply to the relation of painters to their paintings. Analogously, the processes revealed by the sciences are in themselves the action of God as Creator, such that God is not to be found as some kind of *additional* influence or factor added on to the processes of the world God is creating. This perspective continues to be reinforced by the scientific understanding of natural processes, even more strongly than in the recovery of the sense of divine immanence in the eighteenth, nineteenth, and more particularly twentieth centuries, so that I prefer to designate it now as *theistic naturalism.*

It is important to recognize that there are two aspects to God's creative activity: (1) the bringing into existence of the new, in processes whereby novelty and complexity are made to emerge from some prior, earlier, and more basic simpler entities; and (2) the giving of existence to entities other than the Giver. The foregoing account has stressed the renewed emphasis on (1), often denoted as *creatio continua,* under pressure from the cosmological and evolutionary sciences, and this has led to the recovery of the sense of divine immanence. Yet the classical aspect (2) of God's creativity, encapsulated in the dictum of *creatio ex nihilo* and the assertion of contingency (Aquinas' Third Way), is also nuanced by these same new understandings derived from the sciences of the natural world. Since the realization, after Einstein, that time—along with space, matter, and energy—is an aspect of the created order, the notion of "giving existence to" in (2) now has to mean the giving of existence to a *process in time,* such that the distinction between (1) and (2) collapses. We now have to think in terms of *God as Creator continuously giving existence with time to processes* that have the character the sciences unveil; these processes would not go on being and becoming in their particular ways if God were not so continuously giving them such an existence. In other words, God is intimately involved in the created order in both modalities (1) and (2), and the *theistic naturalism* espoused here is intended to incorporate and denote both.

Chapter 4

Panentheism

The scientific picture of the world requires a perspective on God's relation to all natural events, entities, structures, and processes, according to which they are continuously being given existence by God, who thereby expresses in and through them God's own inherent rationality. In principle this should have raised no new problems for Western classical theism when it maintains the ontological distinction between God and the created world. However, it often conceived of God as a necessary "substance" with attributes and with a space "outside" God in which the realm of the created was, as it were, located. Furthermore, one entity cannot exist *in* another and retain its own (ontological) identity if they are regarded as substances. Hence, if God is also so regarded, God can only exert influence "from outside" on events in the world. Such "intervention"—for that is what it would be—raises acute problems in light of our contemporary account of the causal nexus of the world established by the sciences (see Chapter 9). Because of such considerations, this substantival way of speaking has become inadequate in my view and that of many others. It has become increasingly difficult to express the way in which God is present to the world in terms of "substances," which by definition cannot be internally present to each other. This inadequacy of Western classical theism is aggravated by the evolutionary perspective which, as we have just seen, requires that natural processes in the world *as such* need to be regarded as God's creative action.

We therefore need a new model for expressing the closeness of God's presence to finite natural events, entities, structures, and processes and to their very existence; and we need it to be as close as possible without dissolving the necessary distinction between the Creator and what is created. In response to such considerations and those broad developments in the sciences already outlined above which engender a theistic naturalism, there has indeed been a "quiet revolution"[53] in twentieth-century and early twenty-first-century theology involving the resuscitation of the model of pan*en*theism—that admittedly inelegant term for the belief that the Being of God includes and penetrates the whole universe, so that every part of it exists in God and (as against pantheism) that God's Being is more than, and is not exhausted by, the universe.[54] The world is, as it were, "in God," but God is "more than" the world. To say that the world is "in" God evokes a spatial model of the God-world relation, as in Augustine's striking picture of the world as a sponge floating in the infinite sea of God:

> I set before the sight of my spirit the whole creation, whatsoever we can see therein (as sea, earth, air, stars, trees, mortal creatures); yea and whatever in it we do not see. . . and I made one great mass of Thy creation. . . . And this mass I made huge, not as it was (which I could not know), but as I thought convenient, yet every way finite. But Thee, O Lord, I imagined on every part environing and penetrating it, though every way infinite: as if there were a sea, everywhere and on every side, through unmeasured space, one only boundless sea, and it contained within it some sponge, huge, but bounded; that sponge must needs, in all its parts, be filled with that immeasurable sea: so conceived I Thy creation, itself finite, full of Thee, the Infinite; and I said, Behold God, and behold what God hath created; and God is good, yea, most mightily and incomparably better than all these. . . [55]

This "in" metaphor has advantages in this context over the "separate-but-present-to" terminology of divine immanence in Western classical theism. For God is best conceived of as the circumambient Reality enclosing all existing entities, structures, and processes and as operating in and through all, while being more than all. Hence, all that is not God has its existence within God's operation and Being. The infinity of God includes all other finite entities, structures, and processes, for God's infinity comprehends and incorporates *all*. In this model there is no "place outside" the infinite God in which what is created could exist. God creates all-that-is *within* Godself while remaining ontologically distinct.

One pointer to the cogency of a panentheistic interpretation of God's relation to the world is the way the different sciences interrelate to each other and to the world they study. A continuum of sciences from particle physics to ecology and sociology is required to investigate and explicate the embedded systems of natural complexes; a "layered" physicalism is implied, as we saw, for the more complex is constituted of the less complex and all interact and interrelate in systems of systems. It is to this world of multidimensional unity so discovered by the sciences that we have to think of God as relating.

The "external" God of Western classical theism can only be modeled as acting on such a world by intervening separately at the various discrete levels. But if God incorporates both the individual systems and the total System-of-systems within Godself, as in the panentheistic model, then it is more readily conceivable how God could interact with all the complex systems at their own holistic levels. For God is present to the wholes as such as well as to the parts.

Such an emphasis raises the issue of the ultimate transcendence of God as Creator over the contingent natural order. This assertion of God's ultimate transcendence is indeed, I believe, necessary theologically to any understanding of God as Ultimate Reality and Creator, for it entails a view of God as giving existence to all-that-is: entities, structures, and processes. This feature does not arise from the naturalistic presuppositions of this essay but from its theistic stance, namely the need to affirm the reality of God as a transcendent Being giving existence to all else, a consideration properly rooted in the experience of God as Other. It is why the panentheistic model sees God as "more than" the world.

I suggest that considering the nature of human persons can illuminate and help to make clearer what this denotes. For at the terminus of one of the branching lines of natural hierarchies of complexity stands the human person—that complex of the human-brain-in-the-human-body-in-society. Persons can have intentions and purposes, which can be implemented by particular bodily actions. Indeed, the action of the body just *is* the intended action of the person. The physical action is describable, at the bodily level, in terms of the appropriate physiology, anatomy, etc.; but it is also an expression of the intentions and purposes of the person's thinking. They are two modalities of the same psychosomatic event. To be embodied is a necessary condition for persons to have perception, to exert agency, to be free, to participate in community—and to be creative.[56]

In contrast to classical philosophical theism, with its reliance on the concept of necessary substance, the form of panentheism I am espousing here takes embodied personhood as a model of God. It also places a much stronger stress on the immanence of God in, with, and under the events of the world while nonetheless retaining the ultimate transcendence of God, analogously to the way human persons experience their transcendence over their bodies.

Personal agency has been widely used both traditionally in the biblical literature and in contemporary theology as being an appropriate model for God's action in the world. Our intentions and purposes seem to transcend our bodies, yet in fact they are closely related to brain events and can only be implemented in the world through our bodies. Our bodies are indeed ourselves under one description and from one perspective. In personal agency there is an intimate and essential link between what we intend and what happens to our bodies. "We" as thinking, conscious persons appear to transcend our bodies while nevertheless being immanent in them. Our consciousness is a *prima facie* candidate for consideration as a new emergent reality with causal efficacy on its components (in this case, the neurons of the brain[57]). This "psychosomatic," unified understanding of human personhood reinforces the use of a panentheistic model for God's relation to the world. For according to that model, God is *internally* present to all natural entities, structures, and processes in a way that can be regarded as analogous to the way we as persons are present and act in our bodies. Hence the panentheistic model coheres with the *personal* one as one of the justifiable ways of modelling God's agency in the world.

As with all analogies, models, and metaphors, qualifications (the inevitable "is/is not" of all metaphorical language) are needed lest we draw too complete a parallel between God's relation to the world and our relation as persons to our bodies. The first is that the God who, we are postulating, relates to the world like an agent is also the one who creates it, gives it existence, and infinitely transcends it. Indeed, as we saw, the panentheistic model emphasizes this in its "more than the world." By contrast, *we* do not create our own bodies. The second qualification of the model is that we as human persons are not conscious of most of what goes on in our bodies, for example of autonomous functions such as breathing, digestion, and heart beating, or of the sophisticated workings of the neuro-physiological systems that precede and underlie consciousness. Other events in our bodies are of course conscious and deliberate, as we have just been considering. In humans one must thus distinguish between autonomic processes and conscious intentions, but this distinction can scarcely apply to an omniscient God's relation to the world.

The third qualification of the model is that, in speaking of human agency as similar to the way God interacts with the world, we are not implying that God is "a person," that God is, as it were, the personal whole of reality. The notion is rather that God is more coherently thought of as "at least personal," as "*more than* personal," as inclusive of the personal but not defined or limited by it (recall again the "more than" of panentheism). Perhaps we could even say that God is "suprapersonal" or "transpersonal," for there are some essential aspects to God's nature which cannot be subsumed under the categories applicable to human persons. In my view, the panentheistic model allows one to combine a renewed and stronger emphasis on the immanence of God in the world with God's ultimate transcendence over it. It does so in a way that makes the analogy of personal agency both more pertinent and less susceptible than the Western, externalist model to the distortions that we have had to correct by the qualifications (outlined above) of the model of the world-as-God's-body.

The fact of natural (as distinct from human, moral) evil continues to be a challenge to belief in a benevolent God. In the classical perception of God as transcendent and as existing in a "space" distinct from that of the world, there is an implied detachment from the world in its suffering. This thereby renders the "problem of evil" particularly acute. For God can only do anything about evil by intervening from outside, which provokes the classical dilemma: either God can and will not, or God would but cannot; hence God is either not good or not omnipotent. But an ineliminable hard core of offence remains, especially when encountered directly, and often tragically, in personal experience. For the God of classical theism witnesses, but is not involved *in*, the sufferings of the world—even when closely "present to" and "alongside" them.

When faced with this ubiquity of pain, suffering, and death in the evolution of the living world, we are impelled to infer that God, to be anything like the God who *is* Love in Christian belief, must be understood to be suffering in, with, and under the creative processes of the world—an underlying implication of the Christian eucharist, it is worth noting. Creation is costly *to God*, we conclude. Now, when the natural world, with all its suffering, is panentheistically conceived of as "in God," it follows that the evils of pain, suffering, and death in the world are internal to God's own self. God thereby has experience of the natural. This intimate and actual experience of God must also include all those events that constitute the evil intentions of human beings and their implementation—that is, the "moral evil" of the world of human society. God is creating the world from within and, the world being "in" God, God experiences its sufferings directly as God's own and not from the outside. This perception cannot be developed here given the confines of this Essay, but it is already clear that the panentheistic understanding of God's relation to the world has profound "cruciform" implications for the perennial problem of the presence of evil and suffering in creation.

Chapter 5

The Application of an "Emergentist–Naturalistic–Panentheistic" Perspective

It is clear from the foregoing that the sciences have led during the last few decades to a fundamental reassessment of the nature and history of the world which, as has been argued above, now has to be characterized in terms of an *emergentist monism*. The world is a hierarchy of interlocking complex systems; and it has come to be recognized that these complex systems have a determinative effect, an exercising of causal powers, on their components—a whole-part influence. This in itself implies an attribution of reality to the complexes and to their properties, which undermines any purely reductionist understanding. It suggests instead that the determinative power of complex systems on their components can often be construed as a flow of "information," understood in the usual sense of a pattern-forming influence.[58] Be that as it may, successive states of a complex (and most natural entities *are* complexes) are what they are as a result of a *joint* effect of the previous state of the complexes-as-a-whole (which includes their distinctive interaction with their external environment) and of the properties of the individual components. The wholes and parts are intimately interlocked as regards their properties, and they are so in the very existence that a creator God gives them.

In the Prologue (Chapter 1 above) I stressed how the evolutionary perspective engendered by the sciences led to conceiving the work of God as Creator as manifested at all times in natural processes. Since the processes of nature have now to be seen as *both* natural and divine, a *theistic naturalism* has had to be affirmed, according to which God is at work creating and maintaining the evolving, emergent natural order, rather than being conceived as external to it, occasionally intervening and disrupting its regularities. This involves a much tighter linkage between the divine and the natural, and so of the immanence of the divine, than has prevailed in much, at least Western, Christian theology.

These developments in the understanding of God's relation to the natural, albeit created order, which have arisen from pressure from the new scientific perspectives on the world and its history, are integrated in the model of panentheism. This model also attempts to restore the necessary balance between God's transcendence and God's immanence. It is, in fact, closer to an older tradition within theology than has usually been realized, for the Eastern Christian tradition has long been explicitly panentheistic in holding together God's transcendence and immanence.[59] For example, Gregory Palamas (*c.* 1296–1359 C.E.) made a distinction-in-unity between God's essence and his energies; and Maximos the Confessor (*c.* 580–662 C.E.) regarded the Creator-*Logos* as being characteristically present in each created thing as God's intention for it, as its inner essence (its *logos*), which makes it distinctively itself and draws it toward God.

An overwhelming impression is given by these developments, both in the philosophy of science (*emergentist monism*) and in theology (*theistic naturalism* and *panentheism*), of the world as an interlocking System-of-systems saturated, a theist would have to affirm, with the presence of God shaping patterns at all levels. This enhanced emphasis on divine immanence in natural events suggests the application of the same interpretative concepts to the interaction of {God + nature + humanity}—that is, to the human experience of God, the usual focus of theological discourse. In the following chapters an attempt is made to employ theological language about such essentially spiritual (one hesitates nowadays to say "religious") relationships in an *emergentist monist* manner. This includes the interpretation of natural systems, now conceived as exemplifying the activity of God (*theistic naturalism*) and as being "in" God (*panentheism*). For brevity, we will, in the following, denote this fusion of these horizons as *ENP (Emergentist–Naturalistic–Panentheistic).*

In examining spiritual experiences that require a distinctively theological interpretation, that is, putative "complexes" of {God + nature + humanity}, we might hope in this *ENP* perspective to find aspects that parallel what God is also all the time effecting in natural creative processes, namely:

i.   *ontological* features with respect to the emergence of new kinds of reality requiring distinctive language and concepts and referring to *what is there*;

ii. *causal*[60] features with respect to the description of whole-part *influences* of "higher" levels (God, in these instances)[61] on the "lower" ones and referring to *what is being effected and transmitted* in it (and thereby paralleling the pattern-forming associated with the transmission of information in natural systems); and

iii. *transformative* features with respect to the capacities, properties, and (in the case of human persons) experiences of the "lower" basic components of these complex situations.

Our project now is to examine some of those areas of discourse about theological themes that might be illuminated by the awareness that belief in God as Creator involves recognizing the character of the processes whereby God actually creates new forms, new entities, and new structures. They emerge with new capabilities, requiring distinctive language on our part to distinguish them. If God is now acknowledged to be present in, with, and under these processes, then our theology should be able to discern, if only partially, ontological, causal-influential, and transformative features in the interaction of God, nature, and humanity that are similar to those expressed by the concepts of emergent monism, theistic naturalism, and panentheism—in fact, by the whole *ENP* perspective to which we are impelled by the comprehensiveness of the scientific account of the world.

Chapter 6

Jesus of Nazareth—
A Naturalistic Interpretation

The religious consciousness of humanity has played, and still plays, a crucial role in providing individuals and societies with values and hopes that transcend their times and their commitment to their genetic kin. This "long search" of humanity for "God," variously named, in the major religions has been exercised with extraordinary vitality, ingenuity, and richness of expression. In Western civilization we stand on a road that leads from the merger of two of the trails that started in the axial period of 800–200 B.C.E.—those of Greece, partly transmitted via Rome, and those of ancient Israel, as transmitted and transmuted by 2000 years of Christian interpretation and expansion. This double heritage, now compounded with a scientific culture, has shaped our consciousness. It has made us Westerners the people we are. Hence, when we ask the crucial question of the long search, "What can *we* know of God's meaning for humanity in this, our culture?" we are compelled to recognize that, unique among the formative influences in *our* culture, and uniquely challenging in his person and teaching, there stands the figure of Jesus of Nazareth. The Christian tradition has held that human beings have a potentiality, not yet realized, of being in the image and likeness of God, and that the figure of Jesus Christ poses a basic initiative from God toward the actualization of this potentiality.

Nearly three decades have elapsed since the publication of *The Myth of God Incarnate*[62] and the consequent flurry of books it evinced. A range of "Christologies" continue to represent the beliefs of Christians of various colors, with the more orthodox appealing to the Definition of Chalcedon of 451 C.E. as the basis of the formulation of their faith and as a criterion of orthodoxy. It affirmed that Jesus was "complete in regard to his humanity," that is, "completely human"—indeed "perfect" in the sense of "complete"[63]— fully human, but not necessarily displaying perfection in all conceivable human characteristics. Orthodox assessments of Jesus start here, along with recognizing his special vocation and relation to God. Others, less concerned with conformity, have sought modes of expression that are more dynamic than the Chalcedonian Definition, to describe how they believe "God was in Christ reconciling the world to himself."[64]

Jesus' birth

It is widely recognized that the famous Definition, with its uncompromising "two natures" in "one person" and with its underlying "substance" metaphysic, afforded only boundaries within which Christian discourse was urged to range. Patristic reflection (and later reflection in the sixteenth and seventeenth centuries) on its implication of an "exchange of properties" between the human and the divine (*communicatio idiomatum*) did however recognize the significance of the notion of interrelationship—that between the human and the divine. The paradox implicit in the Chalcedonian Definition has continued to be a theological gadfly: an unavoidable assertion, apparently, but with no satisfying resolution of its inherent paradox.

One has no option but to sift and analyse the historical evidence, seeking ways of steering between the Scylla of a "biblicism" that underestimates the difficulties of this ancient literature from a strange culture and the Charybdis of a scepticism that is excessive in relation to the prevailing canons of historical judgment of such sources. The emphasis of recent New Testament scholarship has, in fact, been on a recovery both of the Jewishness[65] and of the humanity of Jesus—whatever the church later attributed to him as the "Christ" of faith.

However, his complete humanity can readily be called into question if it is believed that he was born as the result of a virginal conception by Mary, his mother, without the involvement of any human father. The historical evidence for this is notoriously weak; it is described only in the Gospels of Matthew and Luke and is not mentioned anywhere else in the New Testament, even when the significance of Jesus of Nazareth is at issue. Hence a naturalistic stance with regard to his origins is entirely justified in arriving at a warrantable Christian faith today.[66] The conclusion of the cautious, and very thorough, Roman Catholic scholar Raymond Brown is worth quoting: "the *scientifically controllable* biblical evidence leaves the question of the historicity of the virginal conception unresolved."[67] This verdict would be regarded as over-cautious by other scholars; thus John Macquarrie[68] affirms that "our historical information is negligible. . . . Apart from scraps of doubtful information, the birth narratives [of Matthew and Luke] are manifestly legendary in character."

Biological science also raises acute questions about the "virginal conception." Since females possess only X chromosomes, conception without a father to provide a Y chromosome could lead only to a female child with two X chromosomes—unless there was some kind of divine *de novo* creation of a Y chromosome in the ovum entering Mary's uterus, for the New Testament narratives, even if taken as historical, never deny, and indeed affirm, a normal gestation period. Even such a magical act would be beset with problems: what genes should the DNA of this Y chromosome possess? Those that give the facial characteristics of Joseph, or, if not, of whom? Thus one can go on piling Ossa on Pelion, one improbability on another. Thus the assumption of the full *biological* humanity of a Jesus possessing the normal set of chromosomes is entirely justified.

But a more general consideration now weighs heavily with me because of its theological import. If Jesus is really to be fully and completely human, all that we now know scientifically about human nature shows that he must share both our evolutionary history and have the same multi-levelled basis for his personhood. And that means he must be not only flesh of our flesh and bone of our bone, but also DNA of our DNA. If he does not then, to use the traditional terms, our salvation is in jeopardy, for "what he has not assumed he has not healed."[69]

Hence it is *theologically* imperative that the birth stories of the virginal conception of Jesus be regarded in the same light as those about Adam and Eve—that is, as mythical and legendary (and beautiful) stories intending to convey non-historical and non-biological truths. In this instance the truth being asserted is that God took the initiative in shaping and creating the person and life of Jesus of Nazareth.

However, the question still presses: "*Who* is this Jesus of Nazareth?"

Jesus' life

In one of the best and most influential historical studies of this matter, *Jesus and Judaism*, E. P. Sanders lists eight features about Jesus' career and its aftermath, excluding sayings, which "can be known beyond doubt" and for which "the evidence is most secure."[70]

They are, roughly in chronological order:

1. Jesus was baptized by John the Baptist.
2. Jesus was a Galilean who preached and healed.
3. Jesus called disciples and spoke of there being twelve.
4. Jesus confined his activity to Israel.[71]
5. Jesus engaged in a controversy about the temple.
6. Jesus was crucified outside Jerusalem by the Roman authorities.

7. After his death Jesus' followers continued as an identifiable movement.
8. At least some Jews persecuted at least parts of the new movement, and it appears that this persecution endured at least to a time near the end of Paul's career.

There is, in fact, considerable convergence between various scholars on the contents of such a list. It cannot be stressed too strongly that in our present age one has at least to begin with such a cool assessment of the historical evidence ([2] in Van Harvey's schema, below) concerning Jesus before interpreting his significance for the relation of God and humanity. Such an assessment has been carefully extended and is cogently exemplified by the "Retrospect: A Short Life of Jesus" that Gerd Theissen and Annette Merz append to their *The Historical Jesus*[72] (extracted in Appendix B below).

In practice, one needs to choose, according to Van A. Harvey,[73] between four levels of the meaning of "Jesus of Nazareth":

1. "Jesus as he really was," which is inaccessible;
2. "the historical Jesus," the one who is now recoverable by historical means, about which there is little consensus beyond that noted above;
3. "the memory impression [or 'perspectival image'] of Jesus," as recorded in our resources and recovered through critical analysis, that is, the earliest witness to Jesus recoverable from the traditions; and
4. "the biblical Christ," which designates "the transformation and alteration of the memory-impression (or perspectival image) under the influence of the theological interpretation of the actual Jesus by the Christian community," including the Pauline and Johannine Christologies.

This theological interpretation includes "the idea of pre-existence, the birth and temptation narratives, many of the miracles, those stories which reflect Old Testament prophecies, the resurrection [but see below] and forty-day traditions, and the ascension."[74] The "memory-impression" or "perspectival image" (3) comprises three elements, according to Harvey: "the content and pattern of his teaching and preaching, the form of his actions, and his crucifixion"; this is the important level for our Christological analysis. For Jesus was remembered in the earliest witness as one who raised and answered "the basic human question of faith. It was this role that made him the paradigm of God's action, for he taught them to think of God as the one whose distinctive action is to awaken faith."[75]

In concord with innumerable other New Testament scholars, Christopher Rowland locates "the main thrust of Jesus' message and work in the proclamation of the imminent reign of God [that is, of the kingdom of God]."[76] With others, he also stresses that Jesus identifies his own presence as a sign of the initiation by God of God's imminent "kingdom." This points, *inter alia*, to Jesus having a special relationship with God, which was intense and intimate. We could say that he was exceptionally "open" to God as finally evidenced by his self-offering love manifest in his suffering and cruel death.

But, one may well ask, isn't this apparently objective, historical starting point called into question by the assertion in the traditions about Jesus that there were acts of his and events associated with him that have a "supernatural" connotation: the supposed "miracles" which cannot be reconciled with any naturalistic perspective?

If by a "miracle" one means an event interpreted as not fully explicable by naturalistic means, then judgment must depend on one's *a priori* attitudes toward the very possibility of such events occurring in principle—and a scientific age is, in my view, properly sceptical. Briefly, I consider[77] that in general the healings and apparent exorcisms give rise to no special difficulties, even for a scientific age, but that the "nature miracles" certainly do so; these latter usually have features that either denote them as pure legend or as stories told with an overload of symbolic meanings—in fact, as *true* "myths"! More pertinent to our theme are the major "miracles"[78] connected with the person of Jesus himself.

Jesus' Resurrection

Central to the evidence concerning the historical Jesus are the accounts in the Gospels of his resurrection—the consequences of which echo throughout the rest of the New Testament and shape its contents. In the New Testament one finds[79] various types of "resurrection" affirmations, variously interpreted, including especially the early kerygmatic (i.e., announcing) formulae ("He is risen") and the later narratives. These latter subdivide into those that center on the appearances of Jesus to his disciples and those other, even later ones that concern the finding of the empty tomb. The evidence for this last group of narratives is less reliable and indeed involves particular theological problems connected with the relevance of such a resurrection to that of Christians subsequently.[80]

It is not at all clear that the narratives concerning the "resurrection," taken at their face value, are sensitive to scientific considerations at all, since the end state, the "risen" Jesus, is not open even to the *kind* of repeatable observations science involves. The only science that might by its very nature have any direct bearing on the evidence of the disciples' experience of the risen Jesus is that of psychology. But the probability that these diverse experiences of different kinds of people were due to a kind of communal hallucination or psychosis is minimal in the light of the variety of these same witnesses. This judgment is further supported by their willingness, and the willingness of those to whom they communicated their experiences, to suffer and die for their belief.

The evidence is thus strong that this was a genuine experience within the consciousness of these witnesses. This does not necessarily imply that the experiences were "merely" psychological, with no reference to reality, as long as they can be shown to form part of a meaningful pattern that requires higher-level autonomous concepts to render it intelligible. Just as theories of biology and the neurosciences cannot be reduced in principle to purely chemical ones, so that (as we saw above) realities not present in the separated chemical components have to be depicted as having emerged in these systems, so also a complex of psychological experiences, especially when they are communal, may together manifest a new reality only discernible in that particular complex combination. In the case of the resurrection experiences centered on Jesus, these include the special character of the life, teaching, and death of Jesus; then-current beliefs concerning resurrection; the transformation of the witnesses; and the witnesses' incipient discernment of the presence of God *to* them and so *in* Jesus.

The whole complex of experiences to which the New Testament refers generated this new conceptual framework, the framework of "resurrection," to render these experiences intelligible. This concept or these concepts of "resurrection" thus do not need to be reducible to any purely psychological account. The affirmations of the New Testament that propose this concept can properly be claimed to be referring to a new kind of reality hitherto unknown because not hitherto experienced. "Resurrection" manifests a new kind of ontology in the nature of the risen Jesus.

Such a proposal illustrates the general thesis of this essay: that in theological discourse about experiences of God and of divine action there is a parallel to those processes whereby emergent realities are apprehended in the hierarchy of complex systems studied by the sciences and so are given at least tentative ontological status.[81] In this case the "complex system" in question was the pattern of events constituted by the conjunction of all the special, historical circumstances concerning Jesus, including those of the resurrection experiences. In its totality this conjunction was historical and particular. Given the understanding, described above, of how God interacts with the world and how he can communicate to human beings through meaningful patterns of events,[82] there is every justification for regarding this "resurrection" of Jesus as resulting from the initiative of God and so as a revelation from God manifesting his purposes and intentions at least with respect to Jesus himself. Such a complex of psychological experiences, especially when they are communal, would manifest a new reality only discernible in that particular complex combination.

Thus the concept of resurrection appears not to be reducible to any purely psychological account. The affirmations of the New Testament that propose it can properly be claimed to be referring to a new kind of reality on which the natural sciences as such can make no comment. I recall in closing the penetrating statement of Christopher Evans:

> The core of resurrection faith is that already within the temporal order of existence a new beginning of life from God, and a living of life under God, are possible, and are anticipatory of what human life has in it to be as divine creation; and that this has been made apprehensible and available in the life and death of Christ regarded both as divine illumination of human life and as effective power for overcoming whatever obstructs it.[83]

Chapter 7

Jesus the Christ—
A Naturalistic "Incarnation"?

With all the considerations of the prior chapter in mind, we can now revert to the paradox enshrined in the Definition of Chalcedon concerning the nature of "Jesus the Christ." The sting of that paradox can now be drawn by the recognition of the all-pervasiveness of emergence as a feature of the world. A complex system, as we saw, displays properties that are the *joint* outcomes of the properties of its components and of the system-as-a-whole. This fact allows one to refer to the new reality constituted by the whole without contradicting the particular reality of the components; it also leads one to look for new causal efficacies and transformations. Applying this way of thinking to the person of the historical Jesus leads to the following proposal: the creation of the human personhood of Jesus, born of Mary and fully human, is, according to our total *ENP* perspective, in itself a divine action[84] at all levels, including the biological level, which corresponds to one of the component elements of a "system." At the higher level of the historical Jesus, Jesus' will was fully open in submission to the divine will, such that in his total person there emerged a unique new reality, the God-imbued human being. This emergent reality manifested the divine being insofar as it is expressible in human form—the manifestation of self-offering love, it proves to be, in the light of the life, suffering, and passion of Jesus.

There is, in this *ENP* perspective, no basic contradiction between the biological level of Jesus' existence ("born of Mary," etc.), itself an expression of the divine, and the expression of the divine in the total personhood of Jesus. (By "total personhood" I mean those features of the encounter with him that led his followers to perceive a dimension of transcendence characteristic only of the divine.) The relation between the two levels is parallel to the way God creates new realities by emergence in the natural world. As a result, and also in parallel, the capacities of human nature are transformed by the causal efficacy of the divine presence upon it, such that the risen Jesus becomes the paradigm and paragon of what God intends human beings to be and to become.

Jesus' Jewish followers encountered in him (especially, as we saw, in his resurrection[85]) a dimension of divine transcendence which, as devout monotheists, they attributed to God alone. But they also encountered him as a complete human being, and so experienced an intensity of God's immanence in the world different from anything else in their experience or tradition. Thus it was that the fusion of these two aspects of their awareness—that it was *God* acting in and through Jesus—gave rise to the conviction that something new had appeared in the world that was of immense significance for humanity. A new emergent, a new reality, had appeared within created humanity. Thus it was, too, that they dug deeply into their cultural stock of available images and models[86] (for example, "Christ" = "Messiah" = Anointed, "son of God," "Lord," "Wisdom," "*Logos*"), which were first Hebraic and later Hellenistic, to give expression to this new, non-reducible, distinctive mode of being and becoming instantiated in Jesus the Christ.

When we reflect on the significance of what the early witnesses reported as their experience of Jesus the Christ, we find ourselves implicitly emphasizing both the *continuity* of Jesus with the rest of humanity, and so with the rest of nature within which *Homo sapiens* evolved, and, at the same time, the apparent *discontinuity* constituted by what is distinctive in his relation to God and in what, through him (his teaching, life, death, and resurrection), the early witnesses experienced of God—a discontinuity that was the trigger for his crucifixion (see Appendix B, below). This combination of continuity with apparent discontinuity is just what we have come to recognize in the emergent character of the natural world; thus it seems appropriate to apply this to the cluster of notions concerning the person of Jesus. In Jesus the Christ a new reality has emerged and a new *ontology* is inaugurated. Hence the use of classical imagery, cited above, which was deployed to make explicit the very hiddenness of the self-emptying divine love in the "Christ-event."

This approach prompts a deeper understanding of what is traditionally known as the "incarnation" that occurred in Jesus. This "incarnation" uniquely exemplifies that emergence-from-continuity that characterizes the entire process whereby God is "informing" the world and continuously creating through discontinuity. That is, in light of our understanding of God's creation and presence in the world, we can now interpret "incarnation" not as involving a "descent" into the world by a God conceived of as "above" (and so outside) it—as so many Christmas hymns would have us believe—but as the manifestation of what, or rather of the One who, is already in the world though not recognized or known (which is what the first chapter of John actually says). The human person Jesus is then to be seen, by virtue of his human response and openness to God, as the locus, the *ikon*, in and through whom is made explicit the nature and character of the God who has never ceased to be present, continuously creating and bringing the divine purposes to fruition in the order of energy-matter-space-time. We have to come to see Jesus the Christ as the distinctive manifestation of a possibility always inherently there for human beings by virtue of their potential nature. This makes what he *was* relevant to what we *might be*, for it entails that what we have affirmed about Jesus is not, in principle, impossible for all humanity. Even if, as a matter of contingent historical fact, we think that manifest "incarnation" is only fully to be seen in him, it is not excluded as a possibility for other humans.

The death and resurrection of Jesus, when seen as uniquely manifesting the quality of life that can be taken up by God into the fullness of God's own life, thus implicitly involves an affirmation about what is the basic potentiality of all humanity. It shows us that, regardless of our particular human skills and creativities—indeed regardless of almost all that the social mores of our times applauds—it is through a radical openness to God, a thoroughgoing self-offering love for others and obedience to God, that we can grow into such deep communion with the eternal God that God will not allow biological death to rupture that essentially timeless relation. The new reality of this incarnation has, it transpires, transformative capacities and causal efficacies in taking the humanity of Jesus into the life of God. As a result, it is also potentially transformative of those who follow him in this path of love to God and to "neighbor," their fellow human beings.

How then, in the light of this, might we interpret the experience of God that was mediated to his disciples and to the New Testament church through Jesus? That is, how can we understand the Christ-event as God's self-communication and interaction with the world, such that it is intelligible in light of today's natural and human sciences? It is in these terms that we need to explicate the conclusions of scholars about the understanding in New Testament times of Jesus the Christ. Note, for example, James D. G. Dunn's conclusion that "Initially Christ was thought of . . . as the climactic embodiment of God's power and purpose[:]. . . . God himself reaching out to men . . . God's creative wisdom . . . God's revelatory word . . . God's clearest self-expression, God's last word."[87]

These descriptions of what Jesus the Christ was to those who encountered him and to the early church are all, in their various ways, about God *communicating* to humanity. In the broad sense in which we have been using the terms, they are about an "input of information." This process of "information input" from God conforms with the actual content of human experience, as the conveying of "meaning" from God to humanity.[88] For the conveying of meaning, in the ordinary sense of the word, is implemented initially by a transmission of information, where "information" refers to the constrained and selected elements among all possibilities that sufficiently delimit signals so that they can convey meaning (as in language and other means of human communication).

The use of the concept of "information input" to refer to the way God induces effects in the world was, as far as I know, pioneered by John Bowker:

> It is credibly and conceptually possible to regard Jesus as a wholly God-informed person, who retrieved the theistic inputs coded in . . . brain-processes for the scan of every situation, and for every utterance, verbal and non-verbal. . . . It is possible on this basis to talk about a wholly human figure, without loss or compromise, and to talk also, at exactly the same moment, of a wholly real presence of God so far as that [divine] nature . . . can be mediated to and through . . . the process of brain behavior by which any human being becomes an informed subject—but in this case, perhaps even uniquely, a wholly God-informed subject.[89]

God, in this reading, can convey divine meanings through events and patterns of events in the created world; those under consideration here are the life, teaching, death, and resurrection of the human person, Jesus of Nazareth, as reported by these early witnesses. As the investigations of the New Testament show, the witnesses experienced in Jesus, in his very person and personal history, a communication from God, a revelation of God's meanings for humanity. So it is no wonder that, in the later stages of reflection in the New Testament period, John conflated various concepts in order to say what he intended about the meaning of Jesus the Christ for the early witnesses and their immediate successors. The *locus classicus* of this exposition is, of course, the Prologue to the Fourth gospel.[90] John Macquarrie[91] notes that the expression "Word" or *Logos*, when applied to Jesus, carries multiple undertones: of the image of "Wisdom," a creative intermediary between God and humanity;[92] of the Hebrew idea of the "word of the Lord" for the will of God, expressed in prophetic utterance and in creative activity; and of the Divine *Logos* in Stoicism.[93] This latter concept named the creative principle of rationality and order operative in the universe, especially manifest in the power of human reason, formed within the mind of God and projected into objectivity. Macquarrie suggests substituting "Meaning" for Word-*Logos*, since it helps better to convey the Gospel's affirmation of what happened in creation and in Jesus the Christ. After all, the conveying of meaning (in the ordinary sense) is implemented initially by an input of "information," and this new reality, the incarnation, involves distinctive causal influences.

Human beings can be regarded as consisting of and operating at various levels (the physical, biological, behavioral), which are the foci of the different sciences; these levels merge into that of human culture and its products. The uniqueness of this comprehensive unity-in-diversity and diversity-in-unity that is a human being is encapsulated in our use of the term "person," which denotes both the integrated sum, as it were, over all the discriminable levels and that elusive, almost mysterious nature of the whole. The Christian affirmation, we are suggesting, is that God totally "informed" the human person of Jesus at all levels of his created humanity, and that this "informing" was coincidental and co-ordinate *pari passu* with his total and personal human response of openness and obedience to God his "Father." In this way, Jesus the Christ throws new light on the deeper meaning of the multiple levels of the created world, since these levels were present in him and most of them came into existence through evolution well before the species, *Homo sapiens*, to which he himself belonged. Thus the significance and potentiality of all levels of creation may be said to have been unfolded in Jesus the Christ. In his relation as a created human person to God the Creator, he mediates to us the *meaning* of creation, insofar as we learn through him that for which all things were made as well as how God has been shaping creation for the emergence of persons in communion with Godself. The meaning that God communicates through Jesus the Christ, through the Christ-event, is the meaning of God both *about* humanity and *for* humanity. The meaning he (Jesus) discerns, proclaims, expresses, and reveals is the meaning that he himself is.

Jesus the Christ may therefore be seen as a specific, indeed for Christians a unique, focal point at which the diverse meanings written into the many levels of creation coalesce, like rays of light, with an intensity that so illuminates the purposes of God for us that we are better able to interpret God's meanings communicated in his creative activity over a wider range of human experience of nature and history. This perception of "incarnation" is well-expressed by John Macquarrie:

> Incarnation was not a sudden once-for-all-event . . . but is a process which began with the creation. . . . "Incarnation" . . . is the progressive presencing and self-manifestation of the *Logos* in the physical and historical world. For the Christian, this process reaches its climax in Jesus Christ. . . . The difference between Christ and other agents of the *Logos* is one of degree, not of kind.[94]

What Jesus the Christ was and what happened to him can, in this perspective, be seen as a new source and resource for reading God's meaning for humanity in all the levels of creation that lead to and are incorporated into humanity. It is the clue that points us to a Meaning beyond itself, a key that unlocks the door onto a more ample vista, a focus of rays coming from many directions, a characteristic gesture from the hand of God revealing God's meaning and purpose and nature.

The ideas that generated the *ENP* perspective do indeed seem to illuminate what Christians wish today to affirm about Jesus the Christ as a unique revelation from God about humanity and about God's own self.

Chapter 8

The Eucharist—
A Natural Enactment

The relations of humanity to God may also be illuminated by our understanding of the emergence of new realities in complex, especially self-organizing, systems. For in many of the situations in which God is experienced by human persons one finds complexes of interacting personal entities, material things, and historical circumstances that are not reducible to the concepts that describe their individual components. Could not new realities, and even new experiences of God for humanity, be seen to "emerge" in such complexes and even to be causally effective?

I am thinking,[95] for example, of the church's eucharist (Holy Communion, the Mass, the "Lord's Supper"), in which there exists a distinctive complex of interrelations between its many aspects. Because the eucharist is many-layered in the richness of its meanings and symbols, these aspects can be identified in a wide variety of ways:

1. Individual Christians are motivated by a sense of *obedience* to the traditional re-enactment of the Last Supper, a tradition that from the earliest times in the first century C.E. included the belief that Jesus had authorized the eating of the bread and the drinking of the wine in the same way he had done on that occasion. By re-enacting the Last Supper, Christians identify themselves with Jesus' project in the world.

2. Christians of all denominations have been concerned that their communal act is properly *authorized* as being in continuity with that original act of Jesus and its repetition, recorded in the New Testament, in the first community of Christians. Churches have differed about the character of this authorization but not about its importance.

3. The physical "elements," as they are often called, of bread and wine are, of course, part of the matter of the world and so representative, in this regard, of the created order. So Christians perceive in these actions, in this context and with the words of Jesus in mind, that a *new significance for, and a positive evaluation of, the very stuff of the world* is being expressed in this action.

4. Because it is bread and not corn, wine and not grapes, that are consecrated, this act has come to be experienced also as a new evaluation of the work of *humanity* in *co-creating with God in ordinary work*. As Teilhard de Chardin would put it, in work humanity can be "building the earth," concurring with William Blake: "Let every Christian, as much as in him lies, engage himself openly and publicly before all the World in some Mental pursuit for the Building up of Jerusalem."[96] We can become created co-creators, co-creating creatures.

5. The broken bread and poured-out wine were explicitly linked by Jesus with his anticipated self-sacrificial offering of himself on the cross, in which his body was broken and blood shed to draw all toward unity of human life with God. Christians in this act consciously acknowledge and identify themselves with Jesus' self-sacrifice, thereby offering to reproduce the same *self-emptying love* for others in their own lives and so to further his purposes of bringing in the Reign of God in the world, again—we may say—"for the building up of Jerusalem."

6. They are also aware of the promise of Jesus to be present again in their recalling and remaking of the historical events of his death and resurrection. This "making present" (*anamnesis*) of Jesus, who is regarded as now fully in the presence of—and is, in some sense, identified with—God, is a unique and spiritually powerful feature of this communal act.

7. Eucharist witnesses to the *presence of God,* as the transcendent, incarnate, and immanent Creator.

8. The action is to be undertaken in the *community* of the church, which it both forms and strengthens.

Do we not have in the eucharist an exemplification of the emergence of a new kind of reality requiring a distinctive ontology? For what (if one dare so put it) "emerges" in the eucharistic event *in toto* can only be described in special non-reducible terms such as "Real Presence." Moreover, a kind of divine, transformative, determinative influence (or "causality") is, in this *ENP* perspective, operative in the sacramental experience of the participants. Hence a distinctive terminology of causal influences is also required (involving the use of terms such as "sacrifice" and "grace"), for in the sacrament there is an effect both on the individual and on the community that induces distinctively Christian personhood and society (of "being ever deeper incorporated into this body of love"[97]). So it is not surprising that there is a branch of study called "sacramental theology," which explicates this special reality and the human experience and interpretations of it. Since God is present "in, with, and under" this holistic eucharistic event, God may properly be regarded as distinctively acting on the individual and community through it—surely an exemplification of God's non-intervening, but specific, "whole-part" influence on the world in the *ENP* perspective.

Chapter 9

God's Interaction with the World

In a world that consists of the nexus of causes and effects and whole-part influences that is explicated by the sciences, how might God be conceived as influencing particular events, or patterns of events, in the world without interrupting the regularities observed at the various levels studied by the sciences? The naturalistic model I have proposed[98] is based on the recognition that the omniscient God uniquely knows, over all frameworks of reference of time and space, everything that it is possible to know about the state(s) of all-that-is, including the interconnectedness and interdependence of the world's entities, structures, and processes. This model is rooted in a pan*en*theistic (and so *ENP*) perspective, for it conceives the world as, in some sense, being "in" God, who is also "more than" the world. It also follows that the world would be subject to any divine determinative influences that do not involve matter or energy (or forces).

Thus, it is proposed, mediated by such whole-part influences on the world-as-a-whole (as a *System*-of-systems) and thereby on its constituents, God could cause particular events and patterns of events to occur that express God's intentions. These would then be the result of "special divine action," as distinct from the divine holding in existence of all-that-is, insofar as they would not otherwise have happened had God not specifically intended them in a particular context. By analogy with the exercise of whole-part influence in the natural systems already discussed, such a unitive, holistic effect of God on the world could occur without abrogating[99] any of the laws (regularities) that apply to the various levels of the world's constituents. This influence would be distinguished from God's universal creative action in that particular intentions of God for particular patterns of events to occur are thereby effected. *Inter alia*, patterns of events could be intended by God in response to human actions or prayers.

The ontological "interface" at which God must be deemed to be influencing the world is, on this model, that which occurs between God and the totality of the world (that is, all-that-is), and this may be conceived panentheistically as within God's own self. What is transmitted across the "interface" between God and the world, I have suggested,[100] may perhaps best be construed as a pattern-forming influence—like a flow of what has come to be known technically as "information."[101] However, because of the "ontological gap(s)" between God and the world, which must always exist in any theistic model, one has to admit that this attempt at making intelligible what we postulate as being the effect of God is always seen, as it were, from the human side of the ontological boundary. Whether or not this use of the notion of information flow proves helpful in this context, we do need some way of indicating that the effect of God at all levels is that of pattern-shaping in its most general sense. I am encouraged in this kind of exploration by the recognition that the Johannine concept of the *Logos*, the Word of God, may be taken to emphasize God's creative patterning of the world and hence as God's self-expression *in* the world.

On this model, the question arises at what level or levels in the world such divine influences might be coherently conceived as acting. By analogy with the operation of whole-part influence in natural systems, I have in the past suggested that, because the "ontological gap(s)" between the world and God is/are located simply *everywhere* in space and time, God could holistically affect the state of the world (the whole in this context) at *all* levels. Understood in this way, the proposal implies that patterns of events at the physical, biological, human, and even social levels could be influenced by divine intention without abrogating natural regularities at any of these levels, and that this constitutes the basis for "special divine action" in the world, understood as the nexus of webs of causes and effects and whole-part influences that is expounded by the sciences. This proposal concerning special divine action, a topic that has been such a challenge and focus of contemporary attention,[102] is perhaps more acceptable if the whole-part influence of God is understood to operate mainly at the level of the human person, the emergent reality of which can be located at the apex of the systems-based complexities of the world. God would then be thought of as acting in the world in a whole-part manner by influencing[103] human personal experience,[104] an influence that thereby affects events at the physical, biological, and social levels.

These two limiting forms of the proposal of special divine action—God acting at the level of the whole, and God acting at the level of the human person—are not mutually exclusive. However, divine action in a form that is confined to the personal level is less challenged by (has more "traction" with) the general scientific account of the world than when such divine action is proposed to be at *all* levels. At this stage in my thought, I am inclined to postulate divine whole-part influences at all levels, but with an increasing intensity and manifestation of divine intention from the lowest physical levels up to the personal level, where it could be at its most concentrated and most focused. More general theological considerations need to be brought to bear on how to formulate this model of special divine action. One relevant consideration can nonetheless be developed in this context.

The model as described so far has, I hope, a degree of plausibility in that it depends only on an analogy with complex natural systems in general and on the way whole-part influence operates in them. It is, however, clearly too impersonal to do justice to the *personal* character of the content of many (but not all) of the profoundest human experiences of God. So there is little doubt that the analogy needs to be rendered more cogent by recognizing that among natural systems the instance *par excellence* of whole-part influence in a complex system is that of personal agency, an agency of which we have direct experience. Moreover, I could not avoid speaking above of God's "intentions" and implying that, like human persons, God had purposes to be implemented in the world. For if God is going to affect events and patterns of events in the world, we find we cannot avoid attributing personal predicates such as intentions and purposes to God—inadequate and easily misunderstood as they are. So we have to continue to say that, though God is ineffable and ultimately unknowable in essence, yet God has at least some features describable in the language of personal attributes; using personal language to speak of God is less misleading than saying nothing! In the past I have tried to encapsulate this by affirming that God is "at least personal,"[105] recognizing that being personal can hardly be placed on a graded scale.

If this use of language about God is accepted, we can legitimately turn to the exemplification of whole-part influence in the mind-brain-body relation as a resource for modelling God's interaction with the world. When we do so, the cogency of the "personal" as a category for explicating the wholeness of human agency again reasserts itself, and the traditional, indeed biblical, model of God as in some sense a "personal" agent in the world, acting especially on persons, is rehabilitated—but now in the quite different metaphysical, non-dualist framework of the *ENP* perspective, which is itself coherent with the worldview engendered by the sciences.

Chapter 10

Transforming Grace

Almighty God, in Christ you make all things new:
transform the poverty of our nature by the riches of your grace,
and in the renewal of our lives make known your heavenly glory;
through Jesus Christ your Son our Lord,
who is alive and reigns with you, in the unity of the Holy Spirit,
one God, now and for ever.
— Collect for the Second Sunday of Epiphany[106]

God's action in the world, I have urged in the previous chapter, may be conceived naturalistically (in the sense already adopted in this work) in terms of a whole-part, pattern-forming influence on the constituents of the world, including human beings—analogously, it seemed, to the way personal agency operates on our bodies and the surrounding world. A major aspect of God's action in the world, and the subject of discussion since the fourth century of the Christian church, has been the topic of "grace," which has traditionally been defined as "the supernatural assistance of God bestowed upon a rational being with a view to his sanctification."[107] In light of the model developed here, this modality of God's action may now be given a more naturalistic, rather than a *super*naturalistic, interpretation. Before we attempt to do so, it is worth being reminded of the way "grace" has been interpreted, and elaborated, in the tradition.

Divine grace has been said to be operative in both the individual and in the community of the church. In the former it has been said to be effective in the transformation of individuals; and in the latter it has been invoked *inter alia* in relation to the existence and calling of the church, to the apostolic vocation of its ministers, to the effectiveness of the sacraments, and indeed to the whole range of Christian life within the church, which is seen as derived from God's grace—as exemplified in the ubiquitous use of comprehensive blessings that almost invariably include "the grace of our Lord Jesus Christ." Without going into the complex, and often controversial, history of the clarification of this modality of divine action, some inkling of its scope may be obtained by reproducing the distinctions among even a few of its forms, drawing from *The Oxford Dictionary of the Christian Church*:[108]

1. *Habitual or sanctifying grace*: the gift of God inhering in the soul, by which humans are enabled to perform righteous acts. It is held to be normally conveyed in the sacraments.
2. *Actual grace*: a certain motion of the soul, bestowed by God *ad hoc* for the production of some good act. It may exist in the unbaptized.
3. *Prevenient grace*: that form of actual grace which leads men to sanctification before the reception of the sacrament. It is the free gift of God ("gratuitous") and entirely unmerited.

Actual grace includes:

4. *Efficacious grace*: "grace to which free consent is given by the will so that it always produces its effect";[109]
5. *Sufficient grace*: "grace, which in contrast to efficacious grace, does not meet with adequate cooperation on the part of the recipient, and hence fails to achieve the results for which it was bestowed";[110] and
6. *Prevenient grace*: "the species of Actual Grace which, as an illumination or inspiration of the Holy Spirit, precedes the free determination of the will. It is held to mark the beginning of all activity leading to justification, which cannot be achieved without it, but its acceptance or rejection depends on man's free choice."[111]

And even this list does not exhaust the subtleties distinguishable by theologians![112]

These difficulties, indeed paradoxes, arise because Christian theologians have striven to hold together two propositions: that grace involves God giving Godself to human beings so that they can know and love God and thereby enter into an entirely undeserved relationship with their Creator; and that the ability to receive this gracious approach of God is in itself intrinsic to human nature, which has an affinity and aptitude for it—otherwise it would be irrelevant.[113] In all of its aspects, grace is not a thing but rather the transformation of human life, a transforming personal influence.

I suggest that the paradoxes and complications that have arisen in theological reflection on grace are softened, and even avoided altogether, if one pursues the task we have set ourselves of mapping and constructing a naturalistic model for that particular relation between God, humanity, and nature to which the term "grace" refers. The *ENP* perspective on God's special action in the world (of nature + humanity) leads us to affirm that, in general, it is a whole-part influence, most appropriately conceived of as pattern-forming. Reflection on those spiritual experiences of grace that have been the concern of classical, non-naturalistic theology allows one to discern that they too share the ontological, causal, and transformative features characteristic of natural, interacting complexes. In order to delineate the necessary initial disjunctions and discontinuities, we should acknowledge that here the "complex system" consists of the *ontologically* distinct components of God, humanity, and nature. Furthermore, this interacting "system" evidences a new kind of *causality* of a whole-part kind, for when God acts in a way that can be denoted as an expression of divine grace, there are effects on human beings that are unique and distinctive, necessitating the variety of classical descriptions of the modalities of grace that we have noted above. Thus in these particular kinds of "complex systems" there can occur spiritual experiences that include a *transformation* of the human beings involved, as indicated by the experiences that are analysed in the traditional studies of "grace." We should expect these divine actions to come from *within* any given situation, rather than being imposed externally, for in the *ENP* perspective all that happens to humanity or to the rest of the world is itself *in* God.

As hinted above there is a case, in this perspective, for expanding the term "grace" to include all those actions of God that are transformative, not only of humanity but also of the world. From the standpoint of this expanded theology, grace is operative in individuals, in community, and in nature.

Individuals. It has proved to be a common feature of many reported[114] spiritual ("religious") experiences that the individuals undergoing them have the conviction that there is a power that is in some way purposeful and intentional coming from some "other," and that this power, which may or may not be called "God," *makes a difference* to their lives. The individual comes up against something that goes beyond what any other person can do in helping. These experiences involve some sense of transcendence beyond the ordinary and beyond the individual's inherent powers—a power that seems to have lifted them and set them on their feet again, giving them hope. Those recording such experiences may or may not attribute them to God, but in our *ENP* perspective they would indeed appear to be manifestations of the grace of God. There is often something unexpected and unpredictable about such experiences—so much so that this factor has been designated by Gordon Kaufman as the "serendipitous creativity" of the immanent aspect of the created order.[115] He even goes so far as to give the name of "God" to "that creativity, that mystery, which undergirds our human existence in all its complexity and all its diversity."[116]

Gordon Kaufman and Karl Peters, among others, have certainly drawn attention to a significant feature of humanity's spiritual experience of an Ultimate Reality, which others denote by the word "God." In the theology developed here (e.g., Appendix A and *passim*), I am attributing a more explicit reality-reference to such theological terms than Kaufman and Peters do, so that I am attributing such experiences to the *grace of God*. Nevertheless, the stress is the same, whatever the preferred terminology: a recognition of the unexpected, and a transforming influence from a source other than the recipient of this grace—both characteristic of "emergence." The knots in our entangled patterns of experience are unravelled; there is a new "*in*forming" of human lives, which leads, often, to a *trans*forming of them. All of which is to be seen as an operation of God's grace toward human beings.

The recipient of such divine grace is not moved out of the world but, rather, deeper into it, for it operates from within the world, in the *ENP* perspective, and is appropriate to human beings as embodied creatures:

> As a father has compassion on his children,
> so is the Lord merciful to those who fear him.
> For he knows of what we are made;
> *he remembers that we are but dust.*
> Our days are but as grass;
> We flourish as a flower of the field;
> For as soon as the wind goes over it, it is gone,
> and its place shall know it no more.
> But the merciful goodness of the Lord is from of old
> and endures for ever on those who fear him. . . .[117]

The efficaciousness of the Christian sacraments of baptism and the eucharist, appropriately called "the means of grace,"[118] pivot on such embodiment. But in doing so they also exemplify another essential characteristic of the human individual—the embodiment of the person in community .

Community. As indicated above (Chapter 8), the eucharist of the church is by intention inherently and continuously communal. Likewise, baptism is the rite of initiation of the individual into that community—an incorporation, an embodiment into the family of God. As with the eucharist, the rite of baptism evidences the features of a system characterized by emergence, with "nature" now being represented by water (and sometimes oil), along with the other "components," namely, the baptized human individual and God.

According to classical theology, there is a causal influence at work upon the individual that may be characterized as "regeneration," or at least as the beginning of that process; regeneration involves, over the long term, a transformation of that individual as a "member of Christ." It is now possible to see that one does not need the attribution of *super*naturalistic influences in order to articulate what is going on in baptism; instead, one can interpret it naturalistically and coherently as exemplifying the general mode of action of divine creativity. The new emergent reality of the Christian community is being instantiated and embodied in the action of baptism. In it there is initiated a transformative process of both the individual and the community by God's grace, understood as a whole-part influence in the way we have been delineating here. The whole notion of "being church" is thereby illuminated, for *process* now takes precedence over *structure*.

Nature. One of the implications of that "means of grace," the eucharist, is that the action of divine grace in that context affords a new significance for and a positive evaluation of the very stuff of the world (see Chapter 3 above); the same can be argued for baptism with its use of water (and oil). This entirely concurs with an *ENP* perspective on the created world as already *in* God—we might even say that God is, in a particular sense, "incarnate" in the world, that God is embodied in it.[119] As a result it is appropriate, indeed imperative, to consider the action of divine grace not only within individual persons and communities but also within the created world, for all is "in God." In general, as we have seen,[120] God's action on and within the world is analogous to a transfer of information of a whole-part kind. Because God has a pattern-forming influence on the constituents of the world, there exists the possibility of transforming it at many levels, including the level of the human-brain-in-the-human-body-in-society, and thus the level of human thinking. Given this conception, we may properly speak in an *ENP* perspective of God's grace potentially transforming the world at all levels, and in this sense of God being deeply incarnate in it.

It is urgent, in today's context of ecological disaster, to ask about the human role in the God-world relation as depicted here. In reflecting on the eucharist[121] I have already drawn attention to its implication that human beings are created co-creators[122] with God. If we as individuals and as communities so will, we can indeed be co-creating creatures of and with God in the work of integrating the human-made and God-made constituents of the world—that is, in "making it whole," which is what, in its root meaning, "salvation" means. For God is already at work "in, with and under"[123] the natural processes of the world—just as a Beethoven is totally present *qua* composer in his music as we listen to it. God's grace is already operative in the world, in the very natural processes that the sciences unveil. Our technology, based on those sciences, enables us to co-create with God by working with natural processes, and we can properly call upon God to bestow his grace on humanity in this activity. In a sense, God transforms the world by becoming more incarnate, more embedded in it by means of the co-operation of human agents. Note that, once again, in accordance with our program based on an *ENP* perspective, we are recognizing a distinctive "complex" of God-nature-humanity, manifesting a new causal efficacy from (in this case) God on the constituents of the world and on humanity—an influence that can be transformative.

The *ENP* perspective thus represents a hope and an ideal; it has the potentiality to re-direct the thrust of technology, and thus the work of much of humanity. But the cost is that we will have to learn to love "our neighbor" in the form of the natural world, of which we are an inherent part. We will also have to be recipients of that divine grace that can transform individuals into creative community if we are really to become "church" in its most basic sense, as the channel of transformative grace to all of the world, including those human beings not consciously members of it.

A particular role for humanity, and especially the church, is also implicit in this understanding of the universal operation of divine grace. We have argued that God is everywhere and at all times *in* the processes and events of the natural world, which are to be seen as the vehicle and instrument of God's action and as capable of expressing his intentions and purposes—as our bodies are agents of ourselves. This immediately suggests that human beings should respond to nature with a respect of the same kind as they accord to their own bodies and to the bodies of other people, which are the agents of other "selves" and not mere aggregates of flesh, hair, bone, etc. We respect the body of another person because it is the expression of another "self," the mode and arena of his or her agency. In the case of the natural world, then, if it is God who is the agent who is expressed therein, the human attitude to nature should show a similar respect. Here, however, respect is transmuted into reverence at the presence of God in and through the whole of the created order, which thereby takes on a derived sacredness or holiness as the vehicle and instrument of God's own creative action. But this is precisely to say that the world is *sacramental*, to use the traditional term in Christian theology—a term which, through both its etymology and its long-hallowed associations, also conveys the sense of "holy" in the sense of "set aside for God's purposes."

This complex of proper responses of human beings to nature at once suggests that their role may be conceived as that of *priests of creation*, indeed as *ministers of grace*, as a result of whose activity the sacrament of nature is reverenced; and who, because they alone are conscious of God, themselves, and nature, can mediate between insentient nature and God—for a priest is characterized by activity directed toward God on behalf of others. Human beings alone, as far as we can tell, can contemplate and offer the action of the created world to God. But a priest is also active toward others on God's behalf, and in this sense, too, humanity is the priest of creation. Having reflected on God's purposes, human beings can become active toward the created world; they alone can consciously seek to further and fulfill God's purposes within it.

> Of all the creatures both in sea and land
> Onely to Man thou hast made known thy wayes,
> And put the penne alone into his hand,
> And made him Secretarie of thy praise.
>
> Beasts fain would sing; birds dittie to their notes;
> Trees would be tuning on their native lute
> To thy renown; but all their hands and throats
> Are brought to Man, while they are lame and mute.
>
> Man is the world's High Priest: he doth present
> The sacrifice for all; while they below
> Unto the service mutter an assent,
> Such as springs use that fall, and windes that blow[124]

A qualifying theme. The self-emptying, suffering love of God in creation,[125] in contrast to classical doctrines of the impassibility of God the Father, is a theme that is gradually becoming more widely recognized in contemporary Christian theology; some reference to it must be made before leaving these reflections on divine grace. The natural, created order has to be rule-governed, has to be regular in its activities, insofar as that uniformity that makes science and our free-willing action possible requires a lawlike framework. Many events occur in the natural order that are inimical to life, not least to the human life that has evolved in it naturally through the operation of its own rules. Death and suffering are inevitable components of a creative world that is capable of producing life from insentient matter. In the *ENP* perspective God experiences this intimately and directly; God, we have to aver, suffers "in, with and under" God's creation. Creation is thus costly to God, and there is a self-emptying, a *kenosis*, in God's giving existence to the world and continuing to hold it in existence. Any consideration of the operation of transformative, divine grace in the world in general, and in human beings in particular, has to be placed in this kenotic context. We begin and end with the recognition that divine grace is costly to God, that it involves a sacrificial element which has been variously expressed liturgically, theologically, and aesthetically.

Chapter 11

Conclusion

I have proposed that the principles involved in trying to make clear what is special about spiritual ("religious") situations involving {God + nature + humanity} is broadly applicable[126] to many other experiences of theological concern and interest, both historical and contemporary. We have just seen that "grace" may be conceived as a causally influential and transformative effect of God on a human being, which operates when a person comes into an intimate relation with God under the particular circumstances that characterize that individual life. Transformation is a key feature of the effects of grace on that "emergent whole," which in this context is the human person. Thus there is always an ontological aspect, as well as a causal-influential one, to the experience of grace.

These principles may well also be applicable to the understanding of intercessory prayer in which the "complex" under consideration involves more than one person, though one in particular may be the object of the prayer(s). The whole complex of {God + human persons} can be understood as constituting a new kind of reality with new causal-influential capacities that are *sui generis*. Moreover, the way divine action is envisaged in light of these principles also re-habilitates the distinctive theological notion of "divine revelation" in a more convincing way, I would suggest, than does any "supernaturalist" exposition. Some patterns of apparently "natural" events are more revelatory of the divine intentions and purposes than others.

The *ENP* perspective I have been trying to expound, in conjunction with the new sciences of complexity and of self-organization, provides, it seems, a fruitful and illuminating release for theology from the oppression of excessively reductionist interpretations of the hierarchy of the sciences. It offers a way to render theology's deliberations consistent with what we actually know from science about how the world operates, and it helps to make theological language and concepts accessible to the general exchanges that typify the intellectual life of our times—a milieu from which theology has been woefully and misguidedly excluded for too long.

The new insights into the nature of complex systems and the information-bearing capacity of matter-energy have generated a metaphysic of emergentist monism. When emergentist monism is interpreted in a panentheistic and naturalistic–theistic framework, it allows one to see a congruence and continuity between the nature of matter-energy and the experiences that theology seeks to articulate. Within the framework of an emergentist–monist–panentheistic–naturalistic (*ENP*) perspective, one no longer needs to choose *between* categories such as God, the world, matter, energy, and information. Instead, one can hold them all together in a new kind of synthesis that obviates many of the false dichotomies—the sciences versus the humanities, matter versus the spiritual, science versus religion—which have plagued Western culture for too long.

RESPONSES

Arthur Peacocke's Theology of Possibilities

Philip Hefner

In elaborating what he has chosen to denominate "a Christian natural faith," Arthur Peacocke continues to break ground that opens up significant vistas for Christian theologians. It is a privilege to provide commentary on his project—and in the very act to express gratitude to him for it. My reflections will explore this breakthrough. They are offered from the perspective of a Christian theologian who has himself been freed by Peacocke's proposals to seek a "natural" expression of Christian faith.[1]

The Project: Two Givens

The challenge of interpreting the classic Christian faith drives this new work from Arthur Peacocke's pen, just as it has energized his work up to this point in his career. He derives his theological mandate from a clear set of assumptions: that the first Christian statement about God is that God is Creator, and from this "one would expect the created world to reflect in its very nature the purposes of God, its Creator, and how God and God's relation to the created world are best to be articulated."[2] These assumptions are the foundation, but not the whole, of the theological edifice that he constructs. Christology —described as "understanding the significance of the historical Jesus of Nazareth"—towers above the foundation and extends directly to his interpretation of transforming grace, the experience of God's continuing presence in the world.

His strategy for carrying out this mandate is shaped by the judgment that theological interpretation as a whole has not only not taken account of the "understanding of the natural world afforded by contemporary science" but has rather defended the traditional formulations that need radical revising.

Peacocke goes against the stream of these interpretations by fashioning a *naturalistic theology* whose propria he elaborates:

> Theology may be properly deemed to be "naturalistic" insofar as it assumes that the world of nature is real, that science unveils its realities, and that this natural world contains those entities, processes, and structures that are explicable and eventually rendered intelligible through the natural sciences—hence, . . . no "supernatural" entities, no "miracles" that break the laws or regularities of nature discovered by science, no dualisms within the natural world. . . . Indeed, I also follow Hardwick's strategy, which rests on his (and my) "conviction that theologians must take far more seriously the possibility that naturalism provides the true account of our world, and indeed in its materialist or physicalist version."[3]

Peacocke's starting-point is notable for its distinctiveness among those thinkers who designate themselves as naturalistic theologians. Among these thinkers, the most common strategy is to derive religious faith from scientific knowledge, i.e., to describe scientific understandings of the world in ways that suggest the plausibility of theological interpretations. As I understand it, this is not Peacocke's approach. He begins his work by asserting twin givens: the contemporary scientific understanding of the world *and* the classic Christian faith, which for him includes belief as well as liturgical or sacramental worship. He provides no justification for either science or faith other than the givenness of their importance—science provides the normative understanding of the world we live in, and Christians are rooted in a tradition of faith that is equally normative for them. Peacocke's long espousal of "critical realism" expresses his conviction that both scientific and religious terms should be accorded ontological status. These twin givens are starting points for reflection, not conclusions to an argument. His is not a minimal Christian vision. To be sure, he seeks a bond with thinkers of other traditions, as well as with humanist thinkers, particularly in his treatment of God's ongoing presence in human experience as transforming grace. But he also concentrates on the distinctive and central features of Christian faith—Christ and the eucharist. He does not attempt a scientific proof of God; he would rather speak of science as a pathway on which God is encountered, or as a description of nature in which we discern intimations of God. He wishes to make Christian faith accessible to contemporary people; in this effort he does not argue for faith so much as he *unfolds* it.

It is ironic that at the present time Arthur Peacocke's proposal appears to go against the stream of both Christian theology and scientific ideas of nature. Theology seems reluctant to integrate the naturalist dimension into its methods and contents, while science is frequently suspicious of transcendence. On the theological side, we can count on the fingers of one hand the full-scale expositions of the Christian faith that take the current scientific understandings of nature into account in any meaningful detail. The theological repertoire simply does not give priority to such exposition.

On the side of science, one encounters the widespread opinion that nature is not amenable to spirit or transcendence. The consensus in the scientific community seems to be, first of all, that we cannot speak of nature as teleological, i.e., purposeful, and second, that since nature is an ordered continuum that is shaped by finite causes it does not allow for transcendence. Some go so far as to say that not only is materialism "now the dominant thinking among philosophers and scientists," but there are no established alternatives competing with it. "As a result, theoretical work in philosophy and the sciences is constrained by the various conceptions of what materialism entails."[4] In the context of our present scientific knowledge, it is very risky to affirm that nature and our lives possess meaning and purpose in any deep sense.

Against this backdrop, we recognize the audacity of a theologian who designates his work as *naturalistic* and goes on to insist that nature, particularly as understood by science, is capable of transcendence—and more, that it is an ambience in which the presence of God as set forth in the classic Christian faith can be discerned as real. He recognizes that the naturalist stance

> seems implicitly to preclude the divine. This inference is not a necessary consequence of taking account of the scientific knowledge of the world and its development, however, for a *theistic* naturalism may be expounded according to which natural processes, characterized by the laws and regularities discovered by the natural sciences, are themselves actions of God, who continuously gives them existence.[5]

I introduce *irony* into the discussion, because even though Peacocke goes against the stream of current thinking—both scientific and theological—he in fact represents the historic mainstream of Christian faith and theology. That to so many contemporaries—both within the church and outside it—the Christian faith appears to be *anti*-natural or inherently *super*natural belies the core message of the New Testament, as well as the central Christian dogma of the incarnation of God in Jesus. That dogma, which affirms that Christ is "human, like unto us, divine, like unto God," establishes the principle that the entire created order can be a fit vessel for God's presence. Since human nature includes all of nature—physics, chemistry, biology, and psychology—all of nature must also be capable of receiving the presence of God.[6]

This central belief of Christians did indeed go against the stream of the Hellenistic thought world in which Christianity came to maturity. Nowhere is this more clear than in the philosophical language that this culture offered Christians for the intellectual elaboration of their faith. The irony of their situation is in full view in the classic formulation of the incarnation, the Chalcedonian Formulation of 451 C.E. Generations of scholars and more casual readers have recognized that in this statement Hellenistic ideas and language are simply fractured by the Christian attempt to pour the belief in incarnation into their mold. The notion that the finite could be the bearer of the infinite was more than Hellenistic philosophy could dream of. Christian theologians and philosophers devoted centuries to the effort—unsuccessful, we would judge—to reshape the inherited philosophical concepts so that they could carry the incarnational message. The long and meticulously detailed discussions of the "communication of attributes" (*communicatio idiomatum*) in subsequent centuries witness to the seriousness of the belief that nature is bearer of God's grace. These discussions also find their place in the continuing reflection on the presence of divine grace in the bread and wine of the eucharist. The magnitude of the effort expended in these discussions underscores the impasse that renders this theological tradition conceptually unusable for us today—although by no means irrelevant as a witness to faith.

A further testimony that Christian theology never abandoned the effort to elaborate its basic belief that the material order can be the vessel of divine presence and grace can be found in the traditional theological axiom, discernible in theology from the second century onwards: "*gratia praesupponit naturam, non destruit sed conservat et perficit eam.*" This can be translated in two ways: "grace *presupposes* nature; it does not destroy it, but rather conserves and perfects it" or "grace *undergirds* nature. . . . "[7]

In Christian devotion, this incarnational faith is expressed in a sacramental view that attaches to the public ritual of the eucharist and baptism, for Protestants, and to additional rituals for Roman Catholics and some other Christians. When we place it in this context, we conclude that Peacocke's starting point with the twin givens of science and Christian faith is rooted not in some attempt to give birth to a novel theology—although, to be sure, it is strikingly original in our time—nor in an obsessive and defensive posture of apologetics. Rather, his proposal grows out of his education and experience as a contemporary person who happens to be both scientist and Christian: there is no gainsaying the scientific view of the natural world, if one is truly contemporary, and the logic of the classic Christian faith asks for, nay, demands, articulation in the idiom of that scientific view. It seems, therefore, very "natural" for such a person as Arthur Peacocke to construct the theology that he sets before us. His intention is in accord with the perennial Christian desire to elaborate the experience of God-in-the-world. That his elaboration assumes a conceptual form that offers an alternative to the inadequate Hellenistic metaphysics of nature/supernature and substance, and also reinforces existential meaning, lends additional urgency to his proposals.

Our sense of the "natural" quality of this theology is reinforced by the aura of modesty and tranquility—albeit fully possessed of self-confidence—that marks its exposition. Peacocke presents to us an appealing meditative stance in which to relate religion and science. Although rational thinking and scientific explanation characterize the form of this latest Essay, it is informed at its core by *spiritual discernment*. The easy references to George Herbert, Thomas Traherne, Aubrey Moore, and Charles Kingsley, among others, reveal that he stands in a tradition of Anglican thought that is accustomed to bringing theological and pastoral earnestness together with broader learning so as to articulate religious belief and practice in ways that illuminate contemporary life. The section in the current work entitled "Transforming Grace" (Chapter 10) sets this discerning process in place with forcefulness and delicacy. It opens with a prayer from the Anglican Book of Common Worship, which invokes the presence of God that can "transform the poverty of our nature by the riches of your grace." He goes on to speak of "spiritual ('religious') experiences" in which "individuals . . . have the conviction that there is a power that is in some way purposeful and intentional coming from some 'other,' and that this power, which may or may not be called 'God,' *makes a difference* to their lives."[8]

That Peacocke can set his Anglican tradition of theology, prayer, and transforming experience into the context of nature, scientifically understood, is no small accomplishment. This nature is the locus of grace and at the same time a "physical" realm in which "everything can be broken down into whatever physicists deem ultimately to constitute matter/energy."[9] What unfolds in this Essay and in the author's other works is the process of discerning the sacramental depth of the scientific descriptions. Some twenty years ago, he spoke of this process in terms of "intimations of reality," the title of a book in which, at the end, he comes to this insight:

> And so our two paths to reality of science and religion begin to converge as each points to a depth of reality beyond the power of model and metaphor, in which all that is created is embraced in the inner unity of the divine life of the Creator—transcendent, incarnate, and immanent.[10]

The Project: Three Constructions

R. G. Collingwood made the point more than half a century ago that all of our relationships with nature are mediated through our *ideas* of nature—whether those relationships consist of naïve experience or sophisticated scientific knowledge.[11] Others have concurred with this judgment. I have suggested elsewhere that Collingwood's insight is especially relevant—although certainly not *uniquely* so—for "naturalistic theologians," in that they must construct the idea or concept of nature to which they hold themselves accountable.[12] It becomes necessary, in the process of constructing an idea of nature, also to fashion ideas of science and religious faith. It seems to be the case that we must construct concepts of nature and religion that allow compatibility between the two. Further, each thinker has a somewhat different criterion for determining what constitutes compatibility and how to achieve it. Although there may be a certain level of consensus among thinkers concerning such phenomena as nature and science, each brings a distinctive slant to bear as well.

Although we may all agree that nature is something real and that its reality is independent of us and our knowledge, and even though there is a broad base of agreement in the scientific description of nature, despite this important shared base of understanding each thinker tailors that base in ways that are amenable to his or her intentions. Even when there is complete or nearly complete agreement on the idea of nature, when relating their most deeply held values to nature thinkers "tweak" that idea in ways that enable them to negotiate the relationship as they deem appropriate. I am not implying that this tailoring is a distortion or manipulation—not at all. Rather, I mean that even though nature seems to be a solid reality, the independent existence of which no one doubts, *there is no single authoritative understanding of what it means to take nature seriously, of what constitutes a naturalistic philosophy, or of what it means for religious belief to be compatible with nature*. We have no experience or understanding of nature that is not filtered through our ideas of nature. To a significant degree, the genius of individual thinkers lies in their distinctive fashioning of ideas of nature, science, and religion. This is most certainly true in the case of Arthur Peacocke, and the remaining portion of this commentary offers reflection on the way he has carried out these three constructions—the ideas of nature, science, and Christian faith.

The Idea of Nature

Peacocke employs a distinctive concept of nature in his work. The concept relies fully on scientific understandings, and I do not cite distinctiveness to suggest in any way that he minimizes the scientific component. Rather, his distinctiveness lies in the fact that he brings a richer and more complex set of scientific understandings to the fore, and his theological interpretations do indeed hang on this set of understandings.[13] Furthermore, although not without philosophical assumptions, he brings his theology to bear with a philosophical mediation that is modest in comparison to other naturalistic theologians.[14]

The heart of Peacocke's concept of nature is "a picture of the world as consisting of a complex series of levels of organization and matter in which each successive member of the series is a whole constituted of parts preceding it in the series"[15]—which he gleans from both the natural and the human sciences. This picture is constituted of various levels of complexity, in which wholes and parts "can be observed . . . moving . . . up the ladder of complexity" and also "through cosmic history and biological evolutionary history."[16] The focus is on nature as "layered physicality," which he calls "emergent monism." It is *monistic* "in the sense that everything can be broken down into whatever physicists deem to constitute matter/energy"; it is *emergent* in that from this physical base complex wholes emerge, moving upward on a ladder of increasing complexity. This ladder will finally include society, politics, economics, philosophy, and theology.[17] Each step of the ladder is constituted by a complex whole that is comprised of the interrelationships of its parts. He summarizes his position succinctly:

> *Emergentist monism* affirms that natural realities, although basically physical, evidence various levels of complexity with distinctive internal interrelationships between their components such that new properties, and also new realities, emerge in those complexes—in biology in an evolutionary sequence.[18]

There is considerable subtlety in Peacocke's understanding of scientific emergent monism. He acknowledges both bottom-up and "top-down" or "downward" causation, which he prefers to call downward "determinative influences" rather than "causation."[19] The relation of wholes and parts is also important in his view, because both upward and downward causation occur in the interaction of the wholes (the systems) and the parts that comprise it. Downward determinative influences are in fact the influences of the whole upon the part, whereas upward influences are the converse, the part upon the whole.

We might say that this picture of the scientific view of the world is physicalist at its base, but includes the ladder of emergence up to the human sciences and theology; it is constituted by theories of complexity and emergence, both bottom-up and top-down causal influences. Perhaps we could call it an immensely rich physicalist view of nature. It is a complex concept, but at the same time elegant in its simplicity: Nature is "an interlocking System-of-systems,"[20] driven by the engine of emergence, which in turn brings the downward determinative influences into play and the directionality implied in the ladder of complexity.

This approach not only brings a concept of nature into view; it also includes directives for how nature is to be examined, what our gaze should look for in nature. As "an interlocking System-of-systems" driven by emergence, nature is a realm in which new things come into being at every moment. Consequently, our reflection on nature should be directed at three specific features in nature:

i.   *ontological* features with respect to the emergence of new kinds of reality requiring distinctive language and concepts to refer to *what is there*;

ii. *causal*[21] features with respect to the description of whole-part *influences* of "higher" levels[22] . . . on the "lower" ones and referring to *what is being effected and transmitted* in [the world] (. . . the transmission of information in natural systems); and

iii. *transformative* features with respect to the capacities, properties, and (in the case of human persons) the experiences of the 'lower' basic components of these complex situations.[23]

We will have occasion to note at a later point that what I call this "directed gaze" with respect to our reading of nature will become the theological *leitmotif* for Peacocke's interpretation of God's interaction with the world. He describes his theological effort as the "attempt . . . to employ theological language about such essentially spiritual (one hesitates nowadays to say 'religious') relationships in an *emergentist monist* manner, as in the interpretation of natural systems."[24] The ontological, causal, and transformative features of natural systems become, in his theological method, the lineaments of God's presence in the natural world—in his own distinctive terminology, the lineaments of the complex he designates as {God + nature + humanity}.

The Idea of Science

By any standard, Peacocke's concept of nature is rooted in scientific understandings. At the same time, his integrated, complex intertwining of evolution, layered complexity, emergence, and the downward causation associated with whole/part relationships—in short, the idea of nature as System-of-systems with the nuances of the elements just mentioned—is, so far as I am aware, unique among contemporary theologians. He is distinctive in his steadfastness in holding the manifold elements together in a whole; the parts of the system are clearly enumerated, but they are never allowed to fall away from the wholeness of the system. As much of a consensus as there may be on this concept of nature, it is Peacocke's creative *fashioning* or *constructive work* in putting forth the concept that is impressive.

It is at this point that the idea of nature flows together with the idea of sience. This is not surprising since, as Whitehead observed, science is not our experience of nature; it is our thinking about nature.[25] The flow between science and nature is itself a complex one, as Whitehead further suggested: "We must not slip into the fallacy of assuming that we are comparing a given world with given perceptions of it. The physical world is, in some general sense of the term, a deduced concept. Our problem is, in fact, to fit the world to our perceptions, and not our perceptions to the world."[26] This insight is relevant to Peacocke's work to the extent that he brings to bear theories and conclusions that, although they are scientific, cannot be said to be conclusions of "settled science." He has made an investment in certain scientific hypotheses and theories on which the consensus of the scientific community is not yet firm. Emergence is one case in point: although Peacocke himself has worked with this theory thirty years or more, it is a theory that is yet to be worked out in a full sense. The same must be said about the theory of whole-part determinative influence or "downward causality." Here again, scientists have been enunciating such a hypothesis for half a century or more, but there is still widespread skepticism about it. In a sense, Peacocke is predicting the future of science itself. He is probing the possibilities of science, as well as the possibilities of nature. He is a pathbreaker in interpreting science, as well as in interpreting nature and Christian faith. We must admire his audacity and willingness to take such risk. In so doing, he travels in a path that we must all follow, each of us in our way, if we venture to bring science and theology into constructive concord.

The Idea of Christian Faith

Peacocke certainly does not collapse the tension that Whitehead notes between the world and our perceptions of it. Although he may indeed "tweak" both, so as to enable a fit that is appropriate to his project, he maintains the integrity of each, as is commensurate with the critical realism that has marked his approach for so many years. We observe this same subtle balance or dialectic of tweaking and integrity in his idea of Christian faith. The naturalist premise does set certain parameters, such as the application of the same emergentist monist categories—ontological, causal, and transformative—to the interpretation of natural systems and also to the systems he designates as {God + nature + humanity}. It also rules out supernaturalism and any idea of miracle that defies the regularities of nature. Yet Peacocke leaves no essential element of classic Christian faith out of consideration. He certainly espouses a "higher" Christology than any other naturalistic theologian, as well as a classical view of the liturgical sacraments that is seldom found in naturalistic thought; he conceives of God as "at least personal." Most important, in my mind, is his attention to the ongoing creation of newness, the transformation of the present into the future. His focus is on possibilities, the possibilities of the present—whether in natural systems alone or in the {God + nature + humanity} complex. In this focus on possibility he asserts the most significant dimension of nature and also of God and Christian faith. Possibility is also the realm of spirit and spirituality. Psychologist Mihaly Csikszentmihalyi has written: "Spiritual values, spiritual ideas, symbols, beliefs, and instructions for action . . . point to possibilities to which our biological inheritance is not yet sensitive. The sensate deals with what *is*, the spiritual deals with what *could be*."[27] Csikszentmihalyi's insight finds reinforcement in Peacocke's theological interpretation of human spiritual experience as an encounter with that which transforms and thus empowers human possibilities. This is but one instance, of course, of the working of nature as System-of-systems, which brings forth the possibilities in every level of nature.

Tweaking there may be, on both sides of the theology-science equation, as Peacocke unfolds his theological vision in the framework of nature as a System-of-systems, characterized by emergence and whole-part downward determinative processes, but at no point has he overwhelmed the integrity of either side. He has exemplified in his own work—at the level of culture, science, and theology—the promise of nature's fecundity, revealing new possibilities without end.

A Personal Note

I said at the outset that I myself have found Arthur Peacocke's vision to be liberating to me as a theologian. I will elaborate briefly on the impact of his work on mine. I have for most of my theological career intended to work within a naturalist paradigm. Rooted firmly as I am in the mainline tradition of Christian theology, I have tried to steer my interpretations in an empirical, naturalistic direction. I fancied myself as some form of a "Christian naturalist" theologian. After all, I have been fully grasped by the traditional sources for such a stance—as I have outlined in this essay: the incarnation, the axiomatic disposition of *gratia praesupponit naturam, non destruit sed conservat et perficit eam*, the Lutheran *finitum capax infiniti*, the sacramental tradition of the Real Presence "in, with, and under" natural forms, and the earthy theology of another mentor, Joseph Sittler.[28] In recent years, however, I have been pushed to the belief that I do not fit into the paradigm of naturalism—to my regret. I have been moved to this view, first of all, by my exchanges with scientists who not only reject any association of transcendence with science, but also scoff at those who do assert the association. The hegemony of materialist philosophy among scientists, as I referred to it earlier, seems well entrenched, if my experience is any guide. A second force that has pushed me away from naturalism has been the antagonism and even scorn of certain self-styled naturalistic religious thinkers toward traditional religious faith.

I say that I move away from naturalist theology with regret, because such a move itself violates the Christian faith. Archbishop Temple was surely correct when he said that "Christianity is the most avowedly materialist of all the great religions."[29] Such a move also betrays the widespread and deeply felt desire among Christians and others for theological guides who can relate their faith to nature, scientifically understood—as Peacocke has so often reminded us.

It is here that I wish to express my admiration and gratitude to Arthur Peacocke. The Essay he has written for this book offers a different vision of a natural Christian faith and theology—one that is persuasive to theologians rooted in that tradition. Peacocke's type of naturalist theology goes directly counter to the scientific and naturalist positions that I mentioned above. His alternative is by no means an easy one; it is risky, creative, and demanding—precisely because it does not relax the tension between transcendence and nature and because it insists that scientific understandings, far from precluding Christian faith, actually serve to deepen our understanding of that faith. We can aspire to follow in the path that Peacocke has trodden.

Response 2

Some Words in Favor of Reductionism, Pantheism, Theism, and More

Willem B. Drees

Radical to Be Conservative

Let me first express my gratitude for this "essay in interpretation" from Arthur Peacocke, and for the gift of the lifework of this man to us all—that is, to me and to the many others who value the Christian tradition. I deeply value the critical attitude of modernity that has blossomed in the natural sciences and in historical-critical scholarship regarding the texts of the Christian tradition, and that is reflected in his work. If I were to attempt to summarize Arthur's programme in just a few words, I would describe him as someone who is *radical in order to be conservative*. That is, I see him as someone who is intimately aware of the many major challenges to Christian faith in our time and who seeks to present constructive proposals precisely in order that we may pass on those ideas he judges essential.

Arthur Peacocke's agenda is not to preserve faith by rejecting the naturalistic tenor of science-inspired worldviews, as in the disappointingly shallow debates that assume that faith has to be antagonistic to an evolutionary understanding of the world. To the contrary, by exploring modern understandings of reality in depth, Peacocke seeks to uncover opportunities for Christian faith. He does so, not by treating humans as exceptions to the natural order but by understanding how we relate to the rest of material reality; not by seeking weak spots in modern science and using these to introduce a "God-of-the-gaps" but by engaging with the best available insights—and understanding God as the creator of reality as we have come to know it.

Despite my deep affinity to Peacocke's work, I also have some questions regarding his Essay in this volume and warnings against what I take to be possible misinterpretations. It may be that one needs to be more conservative at some points and more radical at others. In particular, I will stress that emergence is equivalent to reductionism, and will speak more positively than Peacocke does of deism, theism, and pantheism (rather than pan-en-theism) and of divine atemporality. I will challenge personhood as a useful analogy for the God-world relationship, arguing that we need to engage even more with historical-critical scholarship and the secular study of religion. Not that I know better, and even less that I know all the right answers,.but these issues seem to me to be some of the stumbling blocks on our "paths from science towards God."

In fact, *Paths from Science towards God* is the title of a book by Arthur Peacocke.[1] This title is misunderstood, in my opinion, if read in terms of natural theology, as if he were seeking evidence for God through the scientific understanding of the world, that is, as if this were an inductive trajectory. Rather, for Peacocke, the trip does not begin with science but with the Christian heritage, the music of the church, and a wide range of human experiences, values, and ideas. The experience with science and the scientifically known world is part of those experiences, and a critical one. Thus the title could have been "Paths *through* Science to God." The hypothetical-deductive character of the project comes out more clearly in the subtitle of his Essay in this volume, "An Essay in Interpretation," for it is an interpretation of the Christian tradition, but also of "all that is" in the light of that tradition.

A Word in Favor of Reductionism

My first point is not a challenge to Peacocke's position *in toto* but an attempt to preempt a potential misunderstanding. Peacocke's Essay so greatly stresses the emergent character of reality that a simple issue might be lost from sight: *emergence and reduction are two words for the same phenomenon.* As more complex phenomena arise out of simpler antecedents, so the scientific project is to find explanations that can reduce the complexities down to their constituent parts.

Looking upward, we may conclude that higher-level phenomena arise out of properly organized lower-level entities. Higher-level phenomena consist of lower-level entities organized in a certain way. Thus Peacocke speaks in the second chapter of his Essay of a layered *physicalism*; he also limits non-reducibility to an *epistemological* consideration, insofar as "the concepts needed to describe and understand . . . are specific to and distinctive of those levels."[2]

It may thus be that *excessively* reductionist explanations are a problem, especially when epistemological limitations are neglected. But reductionism is not in itself the problem. This is something I learned a long time ago from, if I am not mistaken, Arthur Peacocke himself. It has tremendous significance as a reversal of the implications of reductionism: If matter gives rise to humans, this does not degrade humans. Rather, our esteem for matter should be modified. If reductionism is true, matter has the potential to become a Wolfgang Amadeus Mozart, a Gautama Buddha, and a Jesus.

Of course, much more could be written on the complexity of reality and the variety of methods needed to acquire a more or less complete understanding. The point is that there are unpredictable phenomena, for instance in chaotic systems, and that we have scientific theories and mathematical models that describe and explain the emergence of such systems, including the unpredictability of their behavior and the need for concepts appropriate to each particular level. Thus, the claim of emergentist naturalism is that the "wildness" of reality can be understood scientifically as arising out of more basic processes and thus does not count against a thoroughgoing naturalistic view.

If emergence and ontological reduction are not opposed, what then is the significance of emphasizing emergence for a naturalistic Christian faith? First, emphasizing emergence may bring about awareness of the multileveled character of reality and help us understand how complex phenomena may need new concepts, even though they arise out of simpler ones. Part of that awareness may be that categories used for one level do not match those of a higher level—all forms of money are material, for example, but money is not a material category. In philosophical terms, there may be token-token identity without type-type identity. Thus "emergence" is the catch word for a major task for philosophers, who have to clarify relations between phenomena and theories at various levels.

Second, "emergence" brings to an end arguments from incomprehensibility. It locates complex "emergent" phenomena, which are unpredictable and indescribable in terms of lower-level processes, squarely within the naturalistic, monistic worldview. If life arises out of complex chemistry, there is no explanatory need for a special lifeforce or for divine intervention. It thus undermines one of the arguments for design, or at least for special or "intelligent" design of particular objects in reality.

Whether the laws that rule this fruitful reality are designed is a question not thereby addressed. This brings me to my third comment on emergence: Reductionist-emergentist views need not claim that we will have a complete understanding of reality based on an elementary level that is well understood. Emergence may be about the relation between a lower level that is well known and a higher level that in its complexity is difficult to grasp in full detail—e.g., between chemistry and biology. But emergence (or reduction) may also relate a level that is fairly well known (atomic physics) to an underlying level less known (quarks), and from there to a further level down (superstrings?). As it stands today, we do not know the nature of reality "at the bottom." Thus persistent limit questions regarding the nature of reality, and the most fundamental rules that structure this multilevel emergent reality, may co-exist with a full affirmation of the strength of scientific understanding. If seen as design, the design cannot be seen as a separate phenomenon unto itself but may lie in the basic features of the whole structure.

Yet, fourth, references to emergence are misleading, in my opinion, when it is suggested that it would be easier to envisage divine action in the world at the higher levels than at the level of the constituting entities in space and time. Peacocke quotes a response to Charles Darwin's evolutionary theory as making clear that either God is everywhere or nowhere; but he also writes as if God is found more easily in higher-level phenomena than in lower-level ones. The higher level may need a different description, one that requires its own concepts. Among these may even be religious concepts when we have reached the level of human culture. But none of this offers any particular opportunity for referring to a divine role in the process—unless that divine role is itself of the same kind, and thus also composed of the workings of physical objects. It may be helpful here to speak of a transfer of information, as Peacocke does, but any information transfer still remains a physical change in the world.

Peacocke's stance on emergence is especially open to misunderstanding when, in Chapter 1 of his Essay, he introduces David R. Griffin and his plea for a naturalism freed from materialistic assumptions. Griffin, an advocate of Whiteheadian process philosophy, defends an ontology in which the most fundamental entities or events have a mental as well as a physical pole. Such an ontology might make life easier for some versions of religious thought. But it posits sentience and subjectivity at the most fundamental level, and thus implicitly denies their emergence. If these are emergent properties, they don't have to be among the basic ingredients at the most fundamental ontological level. Furthermore, the approach advocated by Griffin is substantially at odds with current science, whereas the disciplinary order that I have described here expresses insights about the layered character of reality, or at least about the basic order of disciplines. Peacocke is not advocating Griffin's view, but neither does he explicitly note that Griffin presents a "scientific naturalism" quite different from his own.[3]

Divine Atemporality and the Body-Mind Metaphor

If emergence and evolution, both with temporal connotations, are features of the world, the question then becomes what their significance for our worldview might be. Peacocke writes early in his Essay that "dynamic models and metaphors of divine creation and creativity become necessary." I am not against dynamic models or metaphors, but I want to challenge the necessity of such a transfer of temporality in the world to attributing temporality to the divine. Let me first quote another passage from the Essay, in which Peacocke distinguishes between (1) continuing creation as a process in time and (2) creation *ex nihilo*:

Since the realization, after Albert Einstein, that time is an aspect of the created order—along with space, matter and energy—the notion of "giving existence to" in (2 [*creation ex nihilo*]) now has to mean the giving of existence to a *process in time* and the distinction between (1) and (2) collapses. We now have to think in terms of *God as Creator continuously giving existence with time to processes* that have the character the sciences unveil and that these processes would not go on being and involving becoming in their particular way if God were not so continuously giving them such an existence. In other words God is intimately involved in the created order in both modalities (1) and (2)—and the *theistic naturalism* espoused here is intended to incorporate and denote both. (20 above)

I concur that, in the context of General Relativity, time is part of the created order. As has been observed by many, this can be understood theologically in the way Augustine proposed, as creation including the creation of time. On this view, there is no way to ascribe time to the creator in himself—and thus, as Augustine considers, no way to ask what the creator was doing before the creation.[4] What happens in Peacocke's essay is precisely the contrary: he ascribes to God a continuous activity (and thus, temporality). God continuously gives existence to the creation, but this is not a giving in and through time, insofar as time is part of the gift.

This may be a matter of minor interest to many, but the underlying issue is whether we can draw conclusions about the nature of the divine on the basis of analogies from creative processes in the natural order. There is a quite significant risk of neglecting the categorial difference between the created and the creator.

This problem arises also when discussing the relationship between God and the world with the help of the analogy of a person and his or her body. Peacocke writes: "the form of panentheism I am espousing here takes embodied personhood for its model of God . . . while retaining the ultimate transcendence of God, analogously to the way human persons experience their transcendence over their bodies" (23 above). The basic analogy thus seems to be: God transcends and relates to the world as a person transcends and relates to his or her body. However, he then writes that "consciousness is a *prima facie* candidate for consideration as a new emergent reality." Thus, consciousness—not a minor aspect of personhood!—is in a sense secondary to material existence. If the analogy is applied here as well, God would be an emergent reality, dependent upon the world for existence. This is obviously not what Peacocke seeks to argue for, as he qualifies the analogy immediately by stressing God as the one who gives existence to the world. That may well be fine, theologically speaking, but it does make the usefulness of the analogy rather questionable.

Words in Favor of Deism, Theism, and Pantheism

In his Essay, Peacocke pleads for panentheism. While the term may remain somewhat unclear, I have no objections to panentheism. Yet I do have objections to the way various alternatives are dismissed or left out of consideration—and thus, to a presentation that makes panentheism appear as almost unavoidable.

At the beginning of Chapter 3 Peacocke briefly discusses *deism*. The definition given is "the belief in the existence of a supreme being who is regarded as the ultimate source of reality and the ground of value but not as intervening in natural and historical processes by way of particular providences, revelations and salvific acts" (quoting *A New Dictionary of Theology*). Among the "cracks in this conceptual edifice," Peacocke mentions changing ideas regarding the age of the Earth, from a few thousand years to a much longer period, and Darwin's proposal for a mechanism for changes in living beings "that led to the ultimate demise of the external, deistic notion of God's creative actions."

Again, it is not that I object to a more immanentist view of God, but I fail to see how the argument works. If the time scale is much longer than biblical stories suggested, and if there is a natural process that brings about the rich variety of living beings, what does that imply for a deistic understanding of God? I don't see a consequential difference when it comes to thinking of the Deity as the ultimate source of reality. Deism may even have been strengthened by the rise of modern science, given the increased understanding of the integrity of natural processes—thus pushing the need for a particular divine contribution to the horizon of ultimate origins or to the more metaphysical level of giving existence, timelessly or continuously, to natural laws and the order in which they operate.

One might even appropriate one of Peacocke's vivid analogies for a deist appreciation of our world. In the Essay he writes:

> A musical analogy may help—when we are listening to a musical work, say, a Beethoven piano sonata, then there are times when we are so deeply absorbed in it that for the moment we are thinking Beethoven's musical thoughts with him. If, however, anyone were to ask at that moment (unseemingly interrupting our concentration!) "Where is Beethoven now?" we would have to reply that Beethoven-as-composer was to be found only in the music itself. Beethoven-as-composer is/was other than the music (he "transcends" it) but his interaction with and communication to us is entirely subsumed in and represented by the music itself—he is immanent in it and we need not, and cannot, look elsewhere to meet him in that creative role. (above, 19-20)

So too, the deist might say, for God. God is present in his works, the world—not because he is still alive and doing something (just as Beethoven is not doing anything, in the ordinary sense of these words), but because he is the originator of this work, which expresses God's idea for the world.

The argument about time as being part of the created order, as discussed above, allows one to modify the deist position into a more theistic one (but still not panentheist). God gives existence to the world at all times, as the ground of its existence; but when God does this, the created order is still categorically distinct (temporal, finite, dependent) from the divine reality (atemporal, infinite, not dependent upon anything beyond God). Of course, if God and the created order are treated as competing substances, this theistic picture becomes inadequate, as Peacocke writes at the beginning of Chapter 4. But if evolutionary processes in the world are "regarded *as such* as God's creative action," as Peacocke believes, the theist may well concur without thereby being won over to the panentheist model. Perhaps traditional models of primary and secondary causality may be another way to articulate God's involvement without abrogating the rich integrity of natural processes.

Pantheism is only referred to once in the Essay, when Peacocke writes that, in contrast to pantheism, panentheism affirms that God's being is more than the universe. However, I don't see any reason to dismiss a Spinozistic philosophy of life, which equates reality (in all its richness) and God. The intimate relationship Peacocke looks for is affirmed by a pantheist. A major weakness of a pantheist view may be in the absence of any answer to limit questions regarding ultimate origins, but neither is a theistic response that refers to God much of an answer. Rather, these are ways to hint at the mystery that lies at the end of all our exploring.

For me, the point of this brief excursion is not to advocate deism, theism, or pantheism over panentheism. My point is that we ought to acknowledge the way our knowledge *underdetermines* our worldview and thus the multiplicity of acceptable interpretations regarding ultimate origins and ultimate reality. Panentheism may be an interesting interpretation of the world and the tradition, but it is just one such interpretation.

About Interpretation in the New Testament and Religious Studies

Arthur Peacocke is one of the few in the modern religion and science dialogue who pays serious attention to historical-critical studies of the Bible. In reflecting upon the significance of Jesus of Nazareth, Peacocke comes to draw in his Essay upon the work of various scholars. I appreciate this greatly but want to ask whether this should not have been done even more radically. Before engaging with biblical studies, however, we may consider why the engagement with biblical studies is deemed significant at all.

Peacocke writes, "One has no option but to sift and analyse the historical evidence" (above, 30). That is, one has no option *if* the historical facts are deemed relevant. There is another option, and that is to leave the historical material *qua* history aside. Perhaps it is not early Christianity that is normative, but something else. Scholars of religious traditions can point to many examples where it is said that the early tradition is normative, whereas in fact the decisive authority lies in a particular theological scheme or confession or a particular ecclesial authority. It is a choice to invest "the historical facts of the earliest period" with prime authority. Given that this is a choice, it might be more adequate to acknowledge our responsibility in the matter. We make normative decisions, even when we ascribe particular significance to the early history of the Jesus movement. The tradition has been formative; whether it is also normative depends on us granting it such status. Peacocke makes the choice (as does much of Protestant Christianity) to focus on Jesus and the earliest responses to him, though not without bringing into the conversation, as a lens through which this early history is seen, some of the later dogmatic decisions and discussions regarding the nature of Christ. Though I value his engagement with Jesus Christ as pivotal to the Christian heritage, I am puzzled as to how this fits into his approach in general.

The main general insight of historical scholarship on early Christianity is that the gospels that have come down to us are *theological* interpretations, that is, they have been written down as interpretations of Jesus by his followers several decades later, who provide their own assessments of the situation. Thus, to examine the significance of the "virgin birth" it may be more useful to explore what similar stories about significant figures were around at the time than to draw upon modern biology. Not that this would have led to a different conclusion on this topic. When it comes to the resurrection narratives, these present a particular view of Jesus, especially to argue for his vindication by God, even beyond his death.

Focusing first on those who told the stories and on their theological agenda seems to me a far more "natural" road for a naturalistic account than the approach chosen by Peacocke, who seems to make here an exception to his general approach, which is more scientifically oriented. "It is not at all clear," he writes in Chapter 6, "that the narratives of the 'resurrection,' taken at their face value, are sensitive to scientific considerations at all, since the end state, the 'risen' Jesus, is not open even to the *kind* of repeatable observations science involves." A minor issue is that it is a problem to use *repeatable* observations as a requirement for scientific consideration, since this would cast most of natural history (geology, evolutionary biology, and cosmology) into question as well.

Peacocke writes that psychology is the only science that might have a bearing on this topic but then only considers (and dismisses) the option of collective hallucination or psychosis. However, anthropologists, historians, scholars of religious, and many others have observed that humans invent traditions, create stories that suggest purported phenomena to explain the present condition, and bring their religious beliefs with them in providing a theological account of what happened in the past. It is not to be forgotten that quite a number of decades lie between the crucifixion and the earliest extant literature speaking of a resurrection.

In the same context Peacocke writes, "The evidence is strong that this was a genuine experience within the consciousness of these witnesses." Agreed. Note, however, that this is *within the consciousness of these witnesses*. One may understand the resurrection story as part of the theological convictions of the early Christian movement. But the sentence may incite some readers to read it as if Peacocke claims that the resurrection itself is a genuine event, one that fits within his general naturalistic, emergentist-monist view. This way of reading the text is supported when one reads a few lines later that the resurrection was "a manifestation of a new kind of ontology in the nature of the risen Jesus." Here there is no reference to those followers anymore.

A problem with such an ontological understanding of the resurrection is that it is not just the psychology of the witnesses that might be brought to bear upon the understanding of the resurrection, but the whole emergentist-reductionist picture of existence. Mental phenomena are dependent upon bodily phenomena; that is, roughly speaking, physical entities are involved in chemical processes, which together structure living organisms, some of which have consciousness and culture. In testimonies regarding the risen Lord we have an explicit denial of the underlying levels (insofar as neither walls nor the disintegration of the material body by death are relevant), yet they still claim personal presence. In my opinion, the religious affirmations Peacocke draws upon, e.g., at the end of Chapter 6 of the Essay, are beautiful; but they speak more of a new reality and divine initiative than the emergentist monism presented elsewhere in the Essay seems able to accommodate.

More generally, I think we in science and religion (and in systematic theology in general) might benefit from more regular conversation with scholars who study religion as a human phenomenon, whether historical or anthropological. Religious studies may make us aware of the multiple levels at which religion and theology function, that is, not only at the level of metaphysics and ontology, but also at the level of human practices and communities. The question then becomes whether concepts, when they function in stories that serve and guide daily life, are adequately addressed when approached as theoretical proposals.

Let me suggest an analogy with the development of scientific theories. In the succession of scientific theories, say from Isaac Newton to Albert Einstein to superstrings, ontologies have changed drastically, even though there has been, and must be, continuity at the level of the observable results. Seeking too much continuity at the level of theories and concepts may be misplaced— the kind of misplaced approach that is typical not only of fundamental-isms but also of some carefully constructed but highly artificial proposals in science-and-theology.[5]

Peacocke's Essay is clearly not fundamentalist. But in his focus on the res-urrection is there not the risk of concentrating on a concept from the past, seeking to preserve the image, when a more radical reinterpretation might be more fruitful?

A Few Concluding Comments, for the Time Being

There is much that I wholeheartedly agree with in Peacocke's Essay. The number of comments I have made challenging the Essay, or at least some potential interpretations of it, may suggest otherwise; but that is because it is more interesting to articulate disagreements and issue warnings than to underline agreements. I concur with Peacocke on many points: the under-standing of natural reality as multilayered, including an ontology that is basically physicalist and an epistemology that is non-reductionist insofar as higher levels may need their own concepts; the appreciation of science, his-torical scholarship, and critical thinking in general in theology; and the need to avoid a God-of-the-gaps, since current limitations on our knowledge do not provide license to abandon the scientific project but rather stimuli for further research. What we have come to know is religiously at least as relevant as what we do not yet know.

At the end of my *Religion, Science, and Naturalism* I distinguished between wondering humans and wandering humans. *Wondering* humans may engage in speculative religious thought, related to limit questions regarding the ulti-mate nature and ground of reality. It is there that we have to live with our own limitations, with the underdetermination of our views by evidence; we can offer an interpretation (as the Essay does), but not "the answer." *Wandering* humans, facing various existential problems, may draw upon religious nar-ratives and other resources from their religious traditions. I am not so sure how these two fit together—the big ontological questions, and the practical horizon of lived life.

I am most grateful that, in this Essay and in his other work, Arthur Pea-cocke sought to address grand questions and the particular symbols of Chris-tian faith. In the Essay, incarnation and resurrection are central, as is the God-world relationship. It is precisely on those topics that Peacocke's position is most original, but it is also here that the naturalism he advocated is most under stress. I would suggest that a more radically theocentric interpretation of the God-world relationship might avoid many of these problems, as would a more radically naturalistic reinterpretation of Christian faith in terms of existential and axiological (value) roles in life as it is actually lived.[6]

Response 3

Emergence, Naturalism, and Panentheism: An Eastern Christian Perspective

Christopher C. Knight

Over several decades, Arthur Peacocke has attempted, with considerable success, to answer the question of how a theological response to the natural sciences should be formulated. The *ENP* position that he has articulated—Emergentist, Naturalistic, and Panentheistic—provides, in my judgement, the most promising foundation for a future understanding of the relationship between God and the created order. I also believe, however, that there are elements of the superstructure that he has attempted to build on this foundation that are questionable, as well as building materials on hand of which he has yet to make full use.

What Is Theistic Naturalism?

The first issue to which I wish to draw attention involves what it means to speak, as Peacocke does, of a "theistic naturalism." Clearly, as far as his view of God's action as creator is concerned, this term is entirely appropriate, since he sees the processes revealed by the sciences as being "in themselves the action of God as Creator."[1] However, when it comes to what Peacocke calls God's "providential" action, the term "theistic naturalism" can be misleading. For central to his understanding of this providential action is the notion that God must be able, in some sense, to "respond" to events in the world. This leads him to attempt to understand how this might be possible without supernatural intervention. His solution to this problem is a scheme based on what he calls "whole-part constraint."[2]

This scheme does, admittedly, have a strong naturalistic dimension, in much the same way as do other "causal joint" accounts, such as those offered by John Polkinghorne and Philip Clayton.[3] As these two other accounts perhaps make clearer than Peacocke's own does, however, what is envisaged in this kind of scheme is not what we might call a *strong* theistic naturalism, in which the concept of "special" divine action is eschewed. Rather, what is offered in these accounts is simply a more subtle version of special divine action than that offered by the sort of classical supernaturalist understanding in which the laws of nature are seen as being set aside. Instead of coming about through a "setting aside" of this sort, special providence is seen in these accounts as coming about through God's use of natural laws as tools, which remain fully operative at all times. Although divine "intervention" in the old sense is in this way avoided, it is only in a weak sense that the scheme on which these models are based can be said to be "noninterventionist," for it has merely replaced the old interventionist model with another that envisages a more subtle form of divine interference.

It is evident that interference is still envisaged because those who advocate this scheme still, at least implicitly, have a picture of the world in which two outcomes of any situation are in principle possible. One is that which nature, *left to itself* (in the sense of simply being sustained in being) would probably bring about. The other is that which will come about if God, in addition to sustaining the world and its laws, "responds" to events in the world in a *direct, temporal* way. The scheme may not assume the "supernatural" in the old sense of the term, but it still envisages, in any event, a form of "special" divine providence, a new "decision" on God's part which represents a new causal factor in the situation. While the *physical* causes of any event of this kind are, for this model, all natural ones, at another level an additional cause is required to explain why nature, non-deterministic in character, has followed the particular path that it has. What Peacocke rightly perceives as being absent from his view of God's action as creator—a "kind of *additional* influence or factor added on to the processes of the world"[4]—is quite explicitly present in his view of God as the author of providential events.

This means that there is, here, a problem of nomenclature: two distinct frameworks are, by different people, given the label of "theistic naturalism." One is that which I have here called a strong theistic naturalism, in which there is no concept of temporal divine response which brings about events of "special" providence. The other is the sort of weak theistic naturalism, in which—as in Peacocke's understanding—specific divine initiatives, over and above that which gives rise to the world, become a part of any full explanation of events of "special" providence. Given the importance of this distinction, it might be best in future if people who referred to theistic naturalism did so only with a qualifying adjective to indicate which kind they are referring to. (The abbreviations STN for the strong version, and WTN for the weak, might perhaps prove useful in this respect.)

In a similar way, it is also arguable that there are strong and weak types of panentheism. For the panentheistic claim—that everything in the created order is in some sense "in God"—is in fact understood in a number of different ways.[5] For what we might call the "strong" panentheist, there is a distinct tension between the panentheistic position and the sort of "causal joint" language that is, at least in origin, the language of two entities—the agent and that which the agent affects—which are connected only extrinsically. Opponents of panentheism, like John Polkinghorne, must necessarily speak of a causal joint, because for them the central philosophical question about divine action is that of how God "gets into" a world from which he is separated. For the panentheist, however, God doesn't have to "get into" the world because God is, by definition, already there.

The point here is that if we speak of everything as being "in God," in a way that is at least partially analogous to the way in which, for example, my hand is "in me," then causal joint language becomes problematic. Just as the question of how "I" move "my" hand is not easily susceptible to causal joint analysis, so the question of how God does things with the parts of "his" body (which is one way of expressing the panentheistic position) is equally hard to pose in causal joint terms. In both cases, such language seems inappropriate, since it is not the language of a holistic understanding, but one which assumes two separated entities: the agent and that which is outside the agent and which requires manipulation.

At least for what we might call the "strong" panentheist, the laws of nature are not something essentially autonomous, which God must sometimes manipulate if his will is to be effective. They are, rather, a mode of God's presence in the world. (As we shall note presently, the Eastern Christian understanding of the created order has a strong sense of this and is, because of this, an example of strong panentheism.) The "weak" panentheist, by contrast, is one who, while formally disavowing a separation between God and the world, still tends—as Peacocke does in his defence of his whole-part constraint scheme—to pose the problem of divine action in terms that implicitly accept that the laws of nature have an autonomous status that makes them equivalent to something that is "outside" of God. (The language of God's "sustaining" the world and its laws, for example, is a manifestation of this attitude.)

While Peacocke's approach undoubtedly manifests a tendency toward this kind of weak panentheism, however, this tendency seems to be in tension with another factor in his thought. His insistence that God never acts only on some particular part of the world, or only at some particular level of its complexity, seems, in particular, to be a manifestation of an inner orientation that is panentheistic in the strong sense of the term. His analogy between divine action and action within the human body also suggests this, as does his attitude—the significance of which I have noted elsewhere[6]—toward the sacramentality of the created order.

Why does this tension exist? Is it simply that, within the Western philosophical and theological tradition, the strong kind of panentheism is extremely difficult to articulate? Or is it, perhaps, related to the fact that Peacocke's "weak" naturalism seems also to be at odds with his instincts?[7] (A tension of this sort—though expressed in slightly different terms—has, for example, been noted by John Polkinghorne.[8]) My own judgement, from a close reading of all that Peacocke has written, is that both of these factors are at work. What has happened, it seems to me, is that Peacocke has failed to find a way of expressing, in a way that is not deistic, a theological instinct that is inherently both naturalistic and panentheistic in the strong senses of those terms. And because deism is something that he (rightly) finds theologically unacceptable, he has, if only unconsciously, settled on a kind of compromise.

An Alternative Framework

If this is a valid assessment, the question inevitably arises as to whether a non-deistic framework that manifests both strong naturalism and panentheism is possible. My own belief is that it is. The key, I suggest, lies in the traditional thinking of the Eastern Christian world, of which Peacocke has certainly become more aware in recent years, but of which his knowledge is still somewhat superficial. In particular, I suggest that a strong theistic naturalism can—through incorporation into what can be described as a neo-Byzantine model of God's presence and action in the world—offer an essentially new model of divine action.

This model—which I have described elsewhere[9] in greater detail than is possible here—takes its historical bearings from the strand of Greek patristic thinking[10] that culminated in the work of Maximos the Confessor (580–662 c.e.), in which there is a subtle and profound perception of how everything was, in the beginning, created through the divine *Logos* (John 1:1-4). By moulding the philosophical categories available to him to the realities of the Christian revelation as he perceives them, Maximos expresses his faith in terms of the way in which this *Logos* is to be perceived not only in the person of Jesus but also, in some sense, in the "words" (*logoi*) of all prophetic utterance and in the "words" (*logoi*) that represent the underlying principles of all created things from the beginning.[11] Just as the Fourth Gospel's Prologue speaks not of "the sudden arrival of an otherwise absent *Logos*," but rather of "the completion of a process already begun in God's act of creation,"[12] so Maximos too uses the *Logos* concept to describe a continuous process from the beginning of the cosmos to the Christ-event.

The insight that Maximos expresses in this way is, we should note, an explicit manifestation of a more general intuition that is implicit throughout the Byzantine tradition. This is that it is quite wrong to speak of grace as something which is added to "pure nature." Rather, as Vladimir Lossky has noted, the Eastern Christian tradition knows nothing of this "pure nature," since it sees grace as being "implied in the act of creation itself." Because of this, as he goes on to note, the cosmos is seen as inherently "dynamic . . . tending always to its final end."[13]

The general tenor of this approach is something of which Peacocke is aware. What he does not seem to have recognized fully, however, is something at which Lossky hints in the last part of this quotation: the way in which, for Byzantine theology, at least some[14] aspects of divine providence arise from within the creation through the intrinsically teleological factors that have been, so to speak, built into its components. This can be seen with particular clarity in the work of Maximos, in fact, since the *logoi* that constitute the inner reality of created things are, for him, not only uncreated, and as such a manifestation of the divine *Logos* itself. Each characteristic *logos* of this sort is also, as Kallistos Ware puts it, "God's intention for that thing, its inner essence, that which makes it distinctively itself and at the same time draws it toward the divine realm."[15]

This approach posits, then, a model of the created order that is both teleological and Christological. It is a teleological model in the sense that created things are continuously drawn toward their intended final end (though not in a way that subverts human free will and its consequences). It is a Christological model in the sense that this teleological dynamism comes about not through some external created "force" but through the inherent presence of the divine *Logos* in the innermost essence of each created thing.

At the present time, perhaps, few (outside of the Eastern Orthodox tradition) are likely to accept the details of Maximos's philosophical articulation of this model. The reasons for this do not, however, preclude a consideration of what we might call the general "teleological-Christological" character of the vision that he articulates. Indeed, the adoption of a teleological-Christological model of divine action would seem to have several advantages in the context of current debate. Not the least of these is the model's way of envisaging, in its teleological aspect, a mode of divine action that is neither the "special" nor the "general" mode of divine action as these terms are used in Western thinking.[16] By allowing us to transcend the need to make any distinction between what "nature" can do "on its own" and what can only be done through some "special" mode of action, a neo-Byzantine model of this sort would allow us to see God's presence and action in the cosmos simply as two sides of the same coin. In this respect, it would seem not only to tend toward the sort of Western model that speaks in terms of primary and secondary causes but also to provide this model with a far more definitive theological grounding than it has usually been given.

Rethinking Divine Action

The chief problem that arises for this kind of model relates to the question on which Peacocke has focused in his own work: how divine action is related to the workings of a world characterized by obedience to "laws of nature." Here, I would argue, one of the chief difficulties lies not in our belief that the behavior of the cosmos is characterized by such laws but in the particular way in which we usually think about them. This is an issue that can perhaps best be explored in terms of the way in which the strong theistic naturalism of which I have spoken is usually identified, by Western Christians at least, with the deism of the eighteenth century, in which the scope of divine providence is seen as extremely limited.

Here it is important to recognize that this kind of deism is not, from a philosophical perspective, the only kind of strong theistic naturalism that is possible. In itself, such a naturalism assumes nothing more than that the cosmos develops according to "fixed instructions" of a lawlike kind. The possibility that such instructions can bring about subtle and appropriate "responses" to events in the world cannot be precluded in principle. This can be seen through the analogy of human providential action.

Parents' financial support of their children, for example, can be carried out through a standing order to a bank. Such an order can not only include instructions about the transfer of money on a regular basis—the equivalent of ordinary "general" providence as understood by the deists—but can also anticipate specific needs. It can, for instance, include instructions of the kind: "If my daughter provides a receipt for repairs to her car, then transfer to her account—over and above her regular payment—the amount necessary to pay for those repairs." An instruction of this sort has the effect of "special" providence, in that it brings about action in response to a specific rather than a general need—even though it comes about through a "secondary cause" mechanism of the "general" providence kind and no new action on the part of the prime agent is necessary.

This analogy does, of course, have its limitations. In particular, it can be objected that humans cannot anticipate all possible needs, and that an analogy based on a set of "if . . . then" statements can provide neither an elegant model of divine providence nor one based on mechanisms that are conceivable. Neither of these points is, however, strictly relevant. On the first issue, we simply need to note that God's wisdom cannot be thought of as limited in the way that human wisdom is. As to the problem of elegance, it is important to recognize that the analogy I have used is not intended to elucidate the mechanism of divine action, but simply to illustrate an important principle. This is that "responses" of a providential sort can be the result of a "fixed instruction" mechanism.[17] The analogy's mechanism clearly relates more to human limitations than to divine possibilities and, once again, we must remember that God must not be assumed to be limited in the way that we are.

It is, of course, far from easy to understand how God may have set up his providential fixed instructions in a less clumsy way than we humans must. As I have noted elsewhere, however, it is not entirely beyond conjecture.[18] Moreover, the validity of a non-deistic strong theistic naturalism does not depend on the validity of any particular mechanisms that may be suggested. It depends, rather, on an acceptance of the general belief that lies behind the search for such mechanisms: that the creation—with its inbuilt "fixed instructions"—is far more subtle and complex than our present scientific understanding indicates. This may be something that some naturalists will find difficult to accept. It is not, however, incompatible with naturalism as such.

The point here is that the "laws of nature" that can be provisionally identified are those that can be explored through the scientific methodology. Although this methodology may vary somewhat from discipline to discipline, it relies on the repeatability of observation or experiment and on the discernibility of cause and effect. We need to recognize, however, that we cannot preclude the possibility that the cosmos obeys not only the laws that can be identified in this way but also other "fixed instructions" that are not straightforwardly susceptible to this investigative methodology.

Indeed, this is something that may even seem likely when we consider the effects of complexity. For not only are practical repeatability and discernible cause and effect characteristic of only relatively simple systems, which can be effectively isolated from factors that would obscure these characteristics; in addition, important issues related to reductionism in the sciences suggest the necessity of positing laws or organizing principles of a kind that are not susceptible to ordinary scientific investigation but can only be inferred from their general effect.[19] (This, indeed, is at the heart of the "Emergentist" element of Peacocke's proposal.)

This issue of complexity also has important ramifications for our response to anecdotal evidence of "paranormal" phenomena. For there is nothing incoherent in believing that such phenomena may occur through processes which—while following lawlike patterns—are in practice impossible to replicate in a straightforward manner. The failure of laboratory methods in the investigation of such phenomena may simply indicate that they occur only in situations of considerable complexity or extremity.[20] Once this is recognized, the supposed impossibility of paranormal phenomena becomes questionable, and the anecdotal evidence for such phenomena must be given greater weight.

A strong theistic naturalism can be constructed, then, at least in principle, in such a way that the scope of divine action is not limited in the way that the deists assumed. Because of this, the chief theological objection to a strong theistic naturalism is rendered void. There does remain, however, a further objection, which arises from this scheme's apparent implication that God is nothing more than the "absentee landlord" of deistic thought. It is precisely here, however, that a panentheistic framework comes to the rescue of a strong theistic naturalism, since a God who in some sense "contains" the world can hardly be said to be absent from it. For this reason, a strong theistic naturalism will undoubtedly be more persuasive if it is expanded in terms of a panentheistic understanding of the relationship between God and the world, and this persuasiveness will be reinforced if such an expansion is based on something more than an *ad hoc* juxtaposition of the two frameworks.

This is, in fact, one of the reasons that a reworked version of the Byzantine "cosmic vision" suggests itself as a candidate for such expansion. For not only is such a model comparable to a strong theistic naturalism in its rejection of the concept of "special providence," as we have seen; it also arises from a traditional model which, as commentators have noted, constitutes an explicitly panentheistic framework.[21] The possibility of a synthesis of the two frameworks is therefore an intriguing one. It is also, I would argue, a tenable one, provided that we can accept a notion that is intrinsic to the Byzantine understanding of God's action in the world but that has hitherto been ignored by Western advocates of a strong theistic naturalism. This is the notion of teleology.

A Return to Teleology?

The notion of teleology is one that tends to strike a dissonant note in contemporary discussion, since our understanding of the laws of nature was associated historically with the abandonment of the teleological thinking that had characterized the late medieval thinking of the Christian West. Those who dismiss teleology for these historical reasons tend to forget, however, that it was not teleology *per se* that constituted the chief impediment to the development of early modern science. It was, rather, the total philosophical framework then current that had this effect. (Aspects of early modern physics can still, in fact, be expressed in teleological terms.[22]) More importantly, they ignore the fact that we can no longer regard scientific understanding, or the philosophical framework within which that understanding is developed, in the same way as could those of the early modern period who rejected teleological thinking. Not only has contemporary science challenged many of the broad philosophical aspects of early modern science that would tend to negate a teleological understanding.[23] In addition, it has actually evoked questions about teleology in a direct way.

It has, in particular, indicated that a universe whose development depends on laws of nature and on certain fundamental physical constants need not necessarily be a fruitful one of the sort that ours clearly is. Only very particular laws, in fact, together with very "finely tuned" values of those physical constants, provide the possibility of a fruitful universe like our own. This insight—the basis of the so-called anthropic cosmological principle[24]—has been judged by many, including Peacocke, to be, if not persuasive of, then certainly consonant with, the notion of the world's purposeful creation, providing the foundation for a "theology of nature" in which scientific perspectives provide valid insights into the way in which God acts in the world's continuing creation.

If we accept, with those who think in this way, that God's creative action should be understood in naturalistic terms, then we are faced with the question of how we can understand its teleological aspect. Here two key points need to be made. The first is that, in speaking of a teleological factor in this context, we are speaking of something very different in character from the teleological factor assumed in the Aristotelian thought of the late medieval period. It is not a teleological factor that competes with the concept of mathematical laws of nature but rather one that focuses on the meaningful outcome of the working of those laws. It is what we might call a *teleology of complexity*, one which (as in Peacocke's framework) sees significance in the successive emergent properties to which the increasing intricacy of the cosmos' structures gives rise. The second point to be made is that we are not here speaking of the sort of quasi-vitalistic teleology in which the components of the universe are drawn toward an intended final end by some external agent or force. The universe's teleological tendency is, in this scientific perspective, absolutely intrinsic to its components and to the laws that they obey.

The relevance of this second point to the question of divine action becomes clear when we recall the character of the teleological tendency that is posited by the strand of the Byzantine tradition embodied in the work of Maximos the Confessor. For there, too, as we have seen, there is an understanding of the cosmos' teleological tendency which has precisely this intrinsic character.

A recognition of this parallelism cannot, of course, lead in any simplistic way to the claim that the earlier model anticipates an important aspect of contemporary science. Nevertheless, by pointing to the way in which the "laws of nature" perceptible to the scientist have a teleological effect—both in the physical development of the cosmos and in the biological evolution of the species of our planet—scientific perspectives do suggest important parallels between what we now call the laws of nature and what Maximos the Confessor calls the *logoi* of created things. At the very least, it would seem, there is a sense in which, when teleology at this low level is discussed, there need be no dissonance between scientific perspectives and the basic insights of the teleological-Christological model that he articulated.

Further Implications

Moreover, with our current insights into emergentism and the effects of complexity, we can go much further. If we can accept, in the way that I have suggested, that there is no need to limit the universe's "fixed instructions" to scientifically explorable ones, then there is no reason to limit the teleological tendency of created things to the inherent creativity of the particular "laws of nature" that scientists can investigate. Rather, from the perspective of a teleological-Christological model, it is quite possible to see the "laws of nature" that are perceptible to the scientist as representing no more than a "low level" manifestation of what St. Maximos calls the characteristic *logoi* of created things. Over and above such manifestations there may be, at least in principle, higher levels of manifestation which—while still "lawlike" in character—will inevitably lie beyond what the scientific methodology is able to explore.[25]

Thus, I would argue, a teleological-Christological model allows us to acknowledge the general insights about teleology that arise from the natural sciences, and to appropriate these insights in such a way that we can avoid any distinction between general and special providence. The model also allows us to conjecture that, in addition to the kind of low-level teleology that is scientifically observable, there will also be manifestations of a higher-level teleology in such things as "miraculous" occurrences and revelatory experiences.[26] These latter manifestations, while lying beyond what the methods of science can investigate, may still be understood in naturalistic terms and need not in any way be contradicted by a scientific understanding. They can, in principle, account for all that has previously been attributed to God's "special" providence.

Interpreted in this way, a teleological-Christological model of divine presence and providence seems to manifest a number of advantages over competing models of divine action, including Peacocke's own. These are:

1. The model is based on an explicitly theological understanding, rather than on abstract philosophical questions about divine agency.
2. Questions about how God acts "on" the world—as if from outside—are rendered meaningless, since the model rejects the conceptual picture of what the cosmos can do "on its own" or when merely "sustained in being." This means, among other things, that no distinction between "general" and "special" providence can be made, and all aspects of providence are comprehensible in terms of a single, simple model.

3. While the model is strongly naturalistic, there need be no inherent limitation to the scope of divine providence of the sort assumed by deists.

4. The model removes the tension between scientific understanding and belief in divine action because it enables us to incorporate, within a theological perspective, specific aspects of scientific understanding that are sometimes held to challenge religious belief.[27] It also, in an important way, allows the intrinsic limitations of the scientific methodology to be seen with much greater clarity than previously.

In this neo-Byzantine model there is both a connection to Peacocke's understanding and an attempt to go beyond that understanding in a way that seems to accord with his deeper instincts. One of the main differences between us lies, it would seem, not in our intentions or basic theological orientations but in our evaluation of the status of our present scientific knowledge. For Peacocke, though our understanding of the laws of nature is inevitably imperfect, theological understanding is considerably constrained by our present understanding (hence, for example, at least some of his skeptical comments about the notion of Jesus' virginal conception.[28]) For me, however, it is important to recognize that the regularities of the world that can be perceived through the scientist's methodology may not be the only ones that exist.

Indeed, I would go further. Whatever our judgement may be about any particular reported "miracle" (and the reasons for skepticism may often be great), the total religious experience of humankind suggests to me the need to be open to the possibility that there do occur phenomena of the kind usually deemed miraculous. Such occurrences would not, for the model that I have outlined, mean that the logic of the universe's functioning has been violated but simply that this logic cannot be completely comprehended by human logic. (The universe's logical nature is, after all, grounded not in human logic but in the divine *Logos* itself.) There may be laws of nature about which we know nothing scientifically but which are, nevertheless, occasionally significant in their effects.

In its implications for our understanding of divine action, my view here is comparable, in philosophical terms, both to the view of John Polkinghorne, when he speaks of miracles as being analogous to changes of regime in the physical world,[29] and to that of Robert John Russell, who talks of the possibility of a unique event of religious significance being a "first instantiation of a new law of nature."[30] There are, I believe, some laws of nature that can, if only under very unusual circumstances, bring into effect what we can describe, theologically, as a realization of the world's eschatological potential.

This realization is what is sometimes called a "breaking in" of the age to come. For my model, however, this phrase is somewhat unfortunate, since what is envisaged is not a "breaking in" of something that comes from "outside." What is envisaged is, rather, something that the Eastern Christian tradition has often stressed:[31] a "breaking out" of something that is always present in the world, albeit in a way that is usually hidden from us. If this hidden aspect of the world is not susceptible to scientific investigation, this is not because it is not susceptible to a strongly naturalistic understanding. It is simply because its manifestation depends on something that cannot be replicated under laboratory conditions: the faithful response to God of those who recognize him as their creator and redeemer.

Response 4

Empirical Theology
and a "Naturalistic Christian Faith"

Karl E. Peters

I feel privileged to be able to participate in conversation with Arthur Peacocke's "A Naturalistic Christian Faith for the Twenty-First Century." For four decades I have used his work in courses that I've taught in science and religion. Peacocke's careful and humble thinking, his clear writing, his knowledge of various religions and literature, and his exciting metaphors—all make his work interesting for students as well as scholars. Especially helpful is that he seeks an integration of science and Christian thought in a way that takes into account the full spectrum of Christian theology. *Creation and the World of Science* is his foremost example of this.[1] It is one of my favorite books in science and religion, and I'm glad to see that it has been recently republished. "A Naturalistic Christian Faith for the Twenty-First Century" continues to represent all these features of Peacocke's work. He sets his thinking in the context of other kinds of current naturalistic thought, including my own. For this reason I would like in this essay to engage in a conversation regarding the similarities and differences between our respective positions.

The perspective that I bring to this conversation is that of an empirical theologian. Empirical theology is a twentieth-century movement, mainly in American religious thought, that holds that religious ideas must be tested against human experience and thereby justified. Influenced by American pragmatism, empirical theology flourished at the University of Chicago in the first half of the twentieth century, and it continues in the work of members of the Highlands Institute for American Religious and Philosophical Thought and in *The American Journal of Theology and Philosophy*.[2] Because of its methodological empiricism, this kind of theology is usually naturalistic, and many empirical theologians relate their ideas to the contemporary natural and social sciences. My own theistic naturalism is based on empiricism. My hope is that my effort to relate empirical theology to Peacocke's thinking will further the conversation among many people about naturalistic Christian faith and other ways of being religious naturalistically.

What We Have in Common and a Significant Difference

In our theological reflections, Peacocke and I have much in common. We share the idea of grounding our work in contemporary science and using that science in theological reflection. His use of the concept of emergence to develop a philosophical metaphysics of emergent monism calls me to update my own evolutionary theism. In light of the exciting concept of emergence, Peacocke's theistic naturalism calls theology to focus on the immanence of God. This is congruent with the central task of my own theological work—trying to find metaphors and models to help me better understand how God is working in the world. In focusing on divine immanence, we both reject supernaturalism and a substantive view of God. Instead, we try to understand everything, including the divine, in process terms—as dynamic relational systems. At the same time we both wish to preserve the idea of the transcendence of God. Peacocke does this in terms of panentheism, while I remain a theistic naturalist—a difference I'll address below. Yet both of us keep the distinction between creator and created as central in our thought. Finally, although we focus on the idea of creation, we also develop the idea of continual divine creating in relation to human salvation and hence to other theological ideas about God's work in the world. Peacocke has done this much more than I have done, as represented in the concluding half of his Essay. I have recently turned my attention to matters of salvation in a lecture given in the Epic of Creation course at the Zygon Center for Religion and Science.[3] And my book, *Dancing with the Sacred: Evolution, Ecology, and God* explores how evolutionary thinking in science might be understood in a way that helps human beings in their religious living.[4]

An important difference between Peacocke's thinking and mine involves methodology. Forty years ago in graduate school I made an intentional decision to attempt to do empirical theology. Following Charles Sanders Pierce and Henry Nelson Wieman, I decided that everyday experience, refined by the experience of the sciences, would be the methodological touchstone against which to test my theological ideas. Unless I can tie down my concepts about God and salvation to ordinary life experiences, I do not think my ideas carry much weight either in terms of their truth or in terms of how they may provide guidance for my living.[5]

I see some empirical methodological elements in Peacocke's thought. Scientific concepts such as emergence at various levels are subject to empirical tests. Further, he considers experiences associated with several Christian beliefs and practices—experiences of Jesus, of Christian rituals, and of transforming grace—as empirical theologians also do. However, Peacocke does not rely on experience alone to support his theological ideas. If I understand him correctly, the central ideas of the Christian tradition constitute an equally important criterion for doing theology in relation to the sciences.

This leads me to recognize that we live in different religious communities. Peacocke lives in the very rich tradition of Anglican Christianity, where traditional ideas carry considerable authority. I am a Unitarian Universalist. For me Christianity is an important resource, especially since it is the heritage in which I was raised and educated. However, in principle, Christianity has no more authority than any other religion or philosophical outlook. All religious and philosophical ideas, regardless of where they come from, must be judged by how well they help me make sense of the varieties of human experience—the experience against which scientific ideas are tested, the experiences of human beings in general, and my own personal lived experience.

Transcendence in Theistic Naturalism and Panentheism

The difference in our methods leads to a difference in the way Peacocke and I think about transcendence. In a dynamic theistic naturalism there are three ways to conceptualize transcendence.[6] Peacocke and I probably agree on the first two. The first is the transcendence of the future. That which creates the universe is always bringing new, unexpected realities into being. Such transcending can be spoken of as emergence. The second is epistemological transcendence. There always seems to be more to the world than we can articulate with our present concepts. That science is continually making new discoveries and forming new theories is an example of this. Some theologians recognize that our religious ideas are constructs that help organize and interpret our experience but do not exhaust the depths of that experience. Peacocke recognizes this when he suggests that God is personal yet more than personal[7] and when he explores whether God as a pattern-informing influence acts on all levels of existence, on the human level only, or on the different levels to varying degrees.[8] In my own theistic naturalism, I affirm epistemological transcendence with the idea of "creative mystery."[9]

With the idea of creative mystery it is possible to think about a third kind of transcendence—"ontological transcendence." In naturalism we can say that everything is energy-matter and that energy-matter interacts in particular ways to create the various levels of emergent systems. In my naturalistic theism I have modeled this process with ideas from Darwinian evolution and non-equilibrium thermodynamics. However, one can also ask, "How did energy-matter and the process of creative transformation arise in the first place?" Such a question brings me to the limits of naturalism. It leads me to wonder whether there is something more than the space-time universe. Yet to attempt to answer this question would take me beyond empirical, naturalistic theology. So the idea of mystery can express only the possibility of this kind of transcendence.

Peacocke is able to say more about ontological transcendence, which he calls "ultimate transcendence," because, in addition to the empirical sciences, he relies on the authority of Christian tradition. He writes that ultimate transcendence is

> necessary theologically to any understanding of God as Ultimate Reality and Creator, for it entails a view of God as giving existence to all-that-is: entities, structures, and processes. This feature does not arise from the naturalistic presuppositions of this essay but from its theistic stance, namely the need to affirm the reality of God as a transcendent Being giving existence to all else, a consideration properly rooted in the experience of God as Other.[10]

With an allegiance to both naturalism and the Christian theistic tradition, Peacocke is able to move beyond my methodological empiricism and naturalistic theism to a theological position of panentheism. God is both immanent in the world and ontologically transcendent—that is, more than the world.

Two metaphors illustrate the difference between our views. Peacocke uses Augustine's spatial model of the God-world relation; the world is like a "sponge floating in the infinite sea of God."[11] What strikes me about this metaphor and the passage from Augustine that follows is its positive nature. Even though God transcends the world, the metaphor says something positive about the infinite God. The sea that penetrates the sponge is the same sea that is more than the sponge, that transcends the sponge. One can contrast this with another metaphor about the same God-world relation. In the video *Spirit and Nature*, Seyyed Hossein Nasr speaks of Bedouin Arabs sitting around a campfire on a starless night.[12] The campfire illuminates their immediate surroundings. However, beyond the light is impenetrable darkness. Nasr explains that this darkness is not a negative image. While the flickering light of the campfire represents visible reality, the darkness suggests unmanifested reality—God's reality—out of which the world that we can see and know emerges. To me this metaphor indicates that all we can know is the manifest world. The campfire in the darkness suggests that there are limits of our knowing capabilities (epistemological transcendence) and also that there is something more. But what the more is like remains a mystery.[13]

Distinguishing Creator and Created

One of the theological benefits of holding a panentheist view, with its onto-logical transcendence, is that it enables one to distinguish between that which creates the world, God, and everything in the world that is created. Because God is both more than the world and in-and-through the world, Peacocke is able to maintain the creator-created distinction without the ontological dual-ism of natural and supernatural. His panentheism includes the theistic natu-ralist claim that the divine is creatively present within the world, influencing the world to evolve toward more and more complex phenomena. In this lat-ter regard his panentheism also integrates emergent monism into his theol-ogy—into his *ENP* perspective. Theologically, panentheism is the integrating framework, because it incorporates both the ideas of theistic naturalism and emergent monism into a unified theological whole. Its integrating compre-hensiveness makes Peacocke's panentheism a powerful framework for doing theology in relation to the sciences and, in turn, for using science to interpret the major ideas of Christian theology.

Another way to maintain the creator-created distinction is strictly within the framework of theistic naturalism, which does not claim that the creator is more than the spatial-temporal world. Henry Nelson Wieman develops the distinction in relation to a theory of value. In *The Source of Human Good*, Wieman does not think of God as a being who creates the world but as the process of creative transformation.[14] Everything in the world is instrumen-tally or intrinsically good, or both at the same time. Wieman uses the idea of relations of mutual support as a way of generally characterizing what is good. This is a systems view of value. One can apply this understanding of value to emerging systems that evolve in ever greater levels of complexity. In keeping with the basic value claim of Genesis 1, everything in the world can be seen as good because it is an interacting part of a particular whole or a self-sustaining system. And the entire universe as the "system of systems" is very good.

However, from the perspective of Wieman's naturalistic theism there is still something greater than any particular system of sustainable relations of mutual support—no matter how extensive that system is. What is greater is that which creates new relations of mutual support, new systems. Wieman calls this "creative good," and also "creative process," "creative interchange," and "creative transformation." Because it is the source of all intrinsic and instrumental good, Wieman calls this process God.[15] In this way a theological naturalism can make the creator-created distinction without trying to charac-terize God as ontologically transcendent.

Two Metaphors/Models of God as the Source of Ongoing Creation

Metaphors, again, may help clarify a difference between Peacocke's panentheism (which includes theistic naturalism) and my theistic naturalism, which is similar to that of Wieman and also to that of Gordon Kaufman.[16] Like Peacocke, I have found musical metaphors helpful. Thinking of God as the creative process, as those interactions in the world that continually bring the world into being in all its variety, I find it helpful to think of God as the dynamic "music of the spheres"—perhaps like jazz. Jazz is the music created by a system of interacting players. Some provide the beat like the underlying laws of nature. Others, interacting in response to one another, provide the variations on a theme that one begins. Similarly, the interactions within the world produce new natural phenomena, new kinds of species, new kinds of thought, and new ways of living.[17] The persons of a jazz combo in this metaphor represent aspects of the world. They do not represent God. God is the system of interactions among the parts of the world that is creative. "God" is more a verb than a noun.[18]

In "A Naturalistic Christian Faith," Peacocke uses the metaphor of Beethoven and his music. Beethoven is in the music, even as the person Beethoven is more than the music, just as God is in the world even as God is more than the world. This metaphor points to Peacocke's understanding that "God is the *immanent* Creator creating in and through the processes of the natural order. The processes are not themselves God but are the *action* of God-as-Creator—rather in the way that the processes and actions of our bodies as psychosomatic persons express ourselves."[19]

Mind-brain interactions also provide an analogy that helps Peacocke develop his idea that God is working in and through the world but is also more than the world. From the beginning of his section on emergent monism, Peacocke carefully works toward this personal model of God, first by discussing whole-part relationships and downward causation or determinative influences. In a summary of emergent monism he writes that

> the world is a hierarchy of interlocking complex systems; and it has come to be recognized that these complex systems have a determinative effect, an exercising of causal powers, on their components—a whole-part influence. This in itself implies an attribution of reality to the complexes and to their properties, which undermines any purely reductionist understanding. It suggests instead that the determinative power of complex systems on their components can often be construed as a flow of "information," understood in the usual sense of a pattern-forming influence.[20]

Then, from the perspective of theistic naturalism-panentheism, Peacocke suggests that God as the whole (more than the world but in-and-through the world) can be conceived as analogous to an intentional mind that influences the course of the world. God does this by imparting information (Word of God) to shape the patterns that emerge in the world at all levels, especially in human life.[21]

Further, drawing on the authority of Christian tradition, Peacocke holds that this God is love. And he acknowledges that "The fact of natural . . . evil continues to be a challenge to belief in a benevolent God."[22] A benefit, however, of personalistic panentheism is that, because the world is in God, the suffering of the world is experienced by God. Hence, God is a companion in our own suffering, and this exemplifies God as loving. The continual suffering of a loving God stands in contrast to the classical theistic idea that God is detached from the world and its suffering. However, I am led to wonder, if God is experiencing the suffering of the world in Godself, why doesn't God appear to do more to influence the world to relieve unnecessary suffering, just as we humans try to relieve unnecessary and continuous suffering in our own bodies?[23]

In spite of the difference in our theological perspectives—panentheism and theistic naturalism, personal and non-personal theistic models—Peacocke and I recognize that we are trying to conceptualize the same reality that continually creates the world. In comparing the thinking of Kaufman and myself to his own, Peacocke suggests that whatever the preferred terminology is, we all recognize something important that is characteristic of emergence—"the unexpected, and a transforming influence from a source other than the recipient of this grace."[24]

Early in my career, I began thinking of God as "the grace-type event," in which grace means that some new good comes into being beyond human control. I would also say that this understanding of "grace" applies to the natural world. In the interactions between the parts of the world at a particular level of existence, new systems emerge that are not strictly determined by any one part of the system. It is the interactions among the parts and with the external conditions that are the creative source of the new emergent phenomena.

How I Feel Drawn by Peacocke toward Theological Personalism

Even though I suggest a non-personal way of understanding God in my theistic naturalism, I am intrigued by Peacocke's personalistic model of God as shaping events in the world through "pattern-forming influence." What intrigues me is that I think have I experienced such influence. It is not uncommon in my everyday life for things to come together in unexpected yet helpful ways that are beyond my control. Some might say that such events are simply due to chance. I have called them occasions of serendipity. Yet, I wonder if something more is involved—some kind of guiding presence.

I can understand such experiences when I open up my own empiricism in the direction of William James's radical empiricism. Radical empiricism may be compared with what some thinkers call classical empiricism. Classical empiricism focuses on discrete sense perceptions as a way to experience the world and to test ideas about what is experienced. An important part of the scientific method is the appeal to this kind of experience. Discrete perceptions are involved when scientists describe emergent systems, their interacting parts, and how the interactions influence the activity of the parts. Classical empiricism thus methodologically underlies what Peacocke calls emergent monism.

Radical empiricism holds that experience includes not only sense perceptions but also feelings in relation to what is experienced. Following William James, it also holds that our initial experience of something is an experience of a whole, and this includes the experience of the person in relation to the whole. It is thinking that, by analyzing, breaks down the whole experience into discrete sense impressions, particular feelings, and their relationships.

> Perception contains knowledge by acquaintance before it issues in knowledge about an object. Perception includes relations, fringes, and patterns of awareness. While classical empiricists considered the clear and distinct aspects of experience the foundation of knowledge, James considers this vague fringe of relational experience to be more basic to cognition.[25]

The distinction between classical and radical empiricism results in two types of empirical theology, represented by Wieman and Bernard Meland. Nancy Frankenberry sums up the difference as follows:

> Following Henry Nelson Wieman, some empirical theologians restrict the term "knowledge" to that which involves interpretation, reflection, and prediction. Others, after the fashion of Bernard Meland, prefer to widen the term "knowledge" so as to include the mode of acquaintance by which what is directly given is grasped feelingly, and feeling is taken to have cognitive import.[26]

The distinction between these two kinds of empiricism may underlie two ways of looking at some of the experiences Peacocke describes in the latter part of his work, experiences that I'll call "events of grace." One way of looking may view events of grace as the workings of a non-personal creative process—serendipitous creativity, if you will. The other way, the way of radical empiricism, can experience events of grace as the pattern-forming influence of a personal divine intelligence. When I recognize these two ways of experiencing events of grace, I can bring to the fore something that often occurs in my own spiritual experiences—the feeling that there is present in the experiences a guiding presence. Peacocke conceptualizes this aspect of my experience very well when he writes that the activity of God is one of shaping events with pattern-forming influence.

Experiencing Events of Grace and Two Models of God

In the last section of the main part of his Essay Peacocke discusses "transforming grace." From his *ENP* perspective he suggests that the term "grace" can apply to actions of God that transform not only human individuals and communities but also the rest of the world. Grace points to a determinative influence that comes from beyond any particular system that is being transformed. In traditional Christian thought this has usually been defined as "supernatural assistance."[27] However, in light of his theistic naturalism and panentheism, Peacocke is able to provide a naturalistic and personalistic understanding of God as bringing about what I call events of grace.

It seems to me that transforming grace is a more general way of understanding the other Christian themes that Peacocke has developed. He interprets the concepts of resurrection and incarnation, and the practices of the eucharist and baptism, in such a way that they become significant examples of transforming grace in the creation of new emergent systems. Rather than discussing each of these core Christian ideas and practices directly, I will end this conversation with Peacocke by sketching my own recent experience of grace. I think that the two ways in which I will briefly reflect on this experience can be applied to other experiences with which Peacocke works in his application of the *ENP* model.

In early April 2006, my wife, Marj Davis, and I visited our granddaughter Jana, who was in her first year at the University of Mary Washington in Fredericksburg, Virginia. The university campus was lovely, and it was a beautiful spring weekend with cherry trees and other flowers blooming. We enjoyed two of Jana's classes. We had meals and meaningful conversations with her and four of her friends—the kinds of conversations that make an older generation hopeful about the future. We attended a beautiful concert put on by high school and university choirs. And on Sunday morning we drove to Arlington, Virginia, to attend worship at an Episcopal church. We went because Jana's choir was providing some of the music for worship.

The church was an "English style" building—a comfortable house of worship. We sat in the third pew, right behind our granddaughter and her choir-mates in a sanctuary that was almost full. The service was conducted by three priests—two of them women. Some of the liturgy was beautifully sung. But, as a Unitarian Universalist and a naturalistic theist, I was turned off by the theology of the hymns and much of the liturgy, which exemplified a substitutionary theory of atonement. However, the sermon by one of the woman priests, emphasizing love and service, was quite moving. And when our granddaughter and her choir-mates sang their first anthem the pure sound of their harmonious voices was "heavenly." From that point on I experienced a change, a feeling of warmth and love, permeating the "atmosphere."

The feeling reminded me of one that I had years earlier at an ecumenical science and religion conference, where I was the sole Unitarian Universalist. As the outsider I was welcomed enthusiastically. At the end of two days of fruitful discussions, I had attended the closing worship service, the eucharist conducted by the Episcopal clergy. Because I felt that I was in a community of love, I joined my companions in taking communion with them in the presence of love—the presence of Christ.

In Arlington that same feeling of love was present as the priests and congregation began the celebration of the eucharist. One of the priests gave a heartfelt invitation to all present to celebrate. Sitting next to my wife, with whom I do not usually attend church because she is a minister of the United Church of Christ and I am a member of a Unitarian Universalist congregation, I realized that this was a rare opportunity to share in significant religious ritual. I whispered to her, "Let's go up." And we did. We knelt at the communion rail to receive the "body and blood of Christ." Jana and a friend from the choir also came to the rail. And so the three of us, none of whom are Episcopalians, celebrated communion together. In that celebration we grew closer together in love.

Reflecting on this event as a classical empiricist with a non-personal model of God as the creative process, I can see how the various elements that I have described—the family relationships, the beautiful weekend, the choir music, the setting of the service, the way it was conducted, my past experiences, my understanding of God as present when love is present—all came together serendipitously as an event of grace. I can think of the event as an example of serendipitous creativity—of God as the creative process—at work in my life.

Reflecting on this same event as a radical empiricist who attends to feelings as well as perceptions, I also have as part of the experience a "feeling of being led." As the parts of the experience came together, interacting with each other in my mind, I can say that a new event emerged. And a part of the experience of that new event was a feeling of a pattern-forming influence. So, I can understand this event as an example—as one small example in my own life—of what Peacocke describes as an emerging system in which a God, personally understood, is the pattern-forming influence.

Both an impersonal model of God as the creative process and a personal model of God as an intentional pattern-forming influence help me understand and appreciate experiences such as my eucharist experience in Arlington. Depending on whether I attend to the experience as a classical or radical empiricist, I can be either a non-personalistic or a personalistic theistic naturalist. I express my heartfelt thanks to Arthur for helping me see this.

Response 5

Sacrament and Sacrifice:
The Feedback Loops of Religious Community

Donald M. Braxton

> The sacramental principle . . . is by no means confined to any Christian
> interpretation, whether Catholic or Protestant, Western or Eastern. All
> the deepest emotions, experiences and evaluations of human beings
> in every state of culture and in all religions find expression in actions,
> objects and external rites believed to be vehicles conveying spiritual
> benefits to the recipients, so that "inward" and "outward" experience
> meet in a higher unity.[1]

Arthur Peacocke's Essay, "A Naturalistic Christian Faith for the Twenty-First
Century," lays out an ambitious program for the reformulation of Christian
theology congruent with current scientific cosmologies. The core of this program is the embrace of a naturalistic interpretation of basic Christian categories in light of contemporary understandings of emergence in the scientific
disciplines. He calls his position an Emergentist–Naturalistic–Panentheism
(hereafter *ENP*). He summarizes his view thus:

> An overwhelming impression is given by these developments, both in
> the philosophy of science (*emergentist monism*) and in theology (*theistic
> naturalism* and *panentheism*), of the world as an interlocking System-
> of-systems saturated, a theist would have to affirm, with the presence of
> God shaping patterns at all levels.

Peacocke proceeds to apply this cosmological vision to key theological *topoi*
such as the historical Jesus, the doctrine of the incarnation, the eucharist, and
divine action. The end result is a very satisfying theological adventure into the
realms of twenty-first-century theology.

My particular interest in this response essay is to address the specific question of sacramentality, or what is sometimes called the sacramental principle. My own approach to this question is to take a slightly deeper look at the evolution of human community and to situate the peculiar dimensions of Christian community within that context. My proposal is that all human religious communities establish for themselves a twofold feedback system that governs their evolutionary trajectory. I designate the first dimension of this feedback system by the term "sacrament." While my interpretation has a specific history in the Christian tradition, it is hardly limited to that particular cultural expression, as my evolutionary narrative will show. The second dimension of this self-regulatory system I designate by the term "sacrifice." As with sacrament, the idea of sacrifice possesses a significant Christian pedigree, yet it is also by no means limited to that tradition. Sacrament and sacrifice are in effect the cultic and cultural expression of the processes regulating the emergence and maintenance of religious communities. They articulate and channel the positive and negative signals of personal and social well-being in human communities. In the language of the Christian tradition, the Christian community is the in-gathered body of Christ transformed in the death and resurrection, and sustained by the abiding presence of the Holy Spirit. I believe each of these dimensions of Christian community can be rendered intelligible in completely naturalistic terms. Cast against the backdrop of this evolutionary history, Christian communities become once again "available" to those spiritually seeking individuals held at a distance by their disinclination to embrace supernaturalism.

I offer my case in three steps: first, I will outline an evolutionary history of religion in our species with an eye to the specific ritual practices of sacrament and sacrifice. Second, I will present my case for a naturalistic reinterpretation of sacrament and sacrifice in the Christian tradition. Third and finally, I will outline some ethnographic work in the culture of science with an eye to experiential correlates of sacrament and sacrifice. My hope is to offer an analogical bridge between the conceptual worlds of altar and workbench, religious community and scientific team. I suggest these correlations might serve as a framework for future religion and science dialogue rooted in a sacramental perception of reality.

The Evolutionary Context of Religious Community

Every exemplar of the human species carries within itself a four-billion year story of the evolution of life on our planet. From the structure of our cells to the evolution of a central nervous system, from the most primitive reptilian emotions seated in the limbic system to the reproductive strategies of mammals, human beings carry within themselves an entire repertoire of biophysical and behavioral systems that make us what we are. Distinctively human forms of existence must be built upon these pre-existing structures. Evolution tinkers, but it does not reinvent wholesale.

Humans became anatomically modern somewhere between 100,000 and 150,000 years ago, but they did not become behaviorally modern until 50,000 to 30,000 years ago. Most anthropologists explain the transition during this period by reference to a complex rewiring of the human brain.[2] Many factors are likely to have contributed to the modification of the human mind/brain, but certainly the evolution of language was one driving force.[3] Whatever the ultimate mixture of forces driving the change in our species during the Middle/Upper Paleolithic period, the fossil record exhibits a virtual starburst of novel cultural inventiveness around 30,000 years ago. We find, for example, the first real evidence of representational sculpture, primitive symbolic inscriptions, personal ornamentation (beads, pendants, etc.), and the famous cave paintings at such places as Lascaux.[4] Enmeshed in this artwork we see the first physical evidence of what appears to be recognizable religion. Elements of what we think of as religion are likely to have been struggling along in their infancy before this time, of course. For example, we find evidence of burial practices and grave provisioning in older fossil strata, including our Neanderthal cousins, but it is only with artifacts like the painting of the "sorcerer" at Trois-Frères that we find definitive evidence. I believe it is only with the dramatic appearance of the sorcerer and the widespread Venus figurines that we have genuine *evidence* of human beings entertaining clearly religious ideas for the first time.

While religion in some form or another seems to be a cultural universal, it is unlikely that we evolved as a species to be religious. Rather, it seems far more likely that we evolved an entire repertoire of mental aptitudes and discovered that we could also entertain religious ideas as a by-product. The implicated mental aptitudes certainly entail language, social intelligence, and a fundamental interest in, and curiosity about, the natural world. The earliest forms of religion appear to be totemism and anthropomorphism, and they appear to be chiefly concerned with exercising effective control over hunting prospects and facilitating the fertility of the species on which the human group depended. These ends were achieved via the conflation of human and animal identities in a "symbolic communion" of humans, landscape, and animal species. To think of the tribe as a collective embodiment of the lifeforce of a place or an animal is called "totemism." Similarly, to think of animal species or natural forces such as rainfall on which the tribe depended as intentional beings is called "anthropomorphism." These forms of religious thought are both outcomes of the complex rewiring conjectured to have occurred in the prior 50,000-year history. Human beings evolved to be able to think about both human beings and animals in novel, for example religious, ways, even though evolution did not select for this capacity as such. Biologists call such developments "exaptations" to distinguish them from directly selected phenomena (adaptations).

The agricultural evolution of roughly 12,000 years ago marks the next determining stage in the emergence of religious community. Settled existence enabled an increase in the number of humans that could be sustained in a viable community for long periods of time. Whereas foraging tribal units require large swaths of land and seem to optimize at about 150 members, agricultural communities in the Neolithic age could sustain numbers much greater. Aside from shifting the focus of religious concern from the hunted animal to a cereal anthropomorphism, this new development precipitated powerful modifications in the utility of religion to our species.

To understand the change, we need to think about why religion was retained in our species even after it evolved. I already stated that most cognitive scientists and evolutionary psychologists believe that characteristically religious ideation emerged as a by-product of other selected mental aptitudes. In other words, there is no "God gene" or other biophysical adaptation responsible for religion as such. By calling religion an exaptation, evolutionists do not mean that religion cannot be powerfully useful. Religion can certainly contribute to species success or failure even if it remains only mediated via cultural "memes." Once religion became possible in our species, it proved to have a variety of indirect social consequences that made it highly adaptive to prehistoric human beings. The answer turns out to be a numbers game. Religion served to solve what we can call the problem of "moral extension."

Like all the primates, human beings are hierarchical and social creatures. For the vast majority of our species' existence we existed as foraging beings with average group sizes around 150 members. As it turns out, this is also the number of people that human beings can reasonably maintain in memory through day-to-day interactions.[5] This fact is crucial since systems of reciprocal altruism in social units depend upon the frequent dispersal of information about potential cooperation partners. Exceed that number and the social unit experiences dissension and likely fission. Cognitive limits associated with the magic number of 150 serve as a kind of cultural bottleneck to the extension of moral consideration in human social units.

The agricultural revolution made it possible for human beings to exceed this number by sustaining larger populations in smaller and smaller regions. If the social glue of these novel communities was to withstand the pressures of social dissension, it required a new force to do the heavy moral lifting. Into this cultural niche and need stepped the readily at-hand cultural construct of religion. As Shermer notes, "The codification of moral principles out of the psychology of the moral traits evolved as a form of social control to ensure the survival of individuals within groups themselves. Religion was the first social institution to canonize moral principles. . . . "[6]

In many cultures around the world, religion and law mean almost the same thing. Historian of religion Eric Sharpe points out that polyvalent words such as *"dharma," "torah,"* "commandment," and even the Latin root of "religion" (*religio*) itself, are interchangeably translatable in legal or religious terms.[7] By interjecting into human community thoughts about moralizing, highly powerful, and very well informed supernatural agencies, it was possible to extend the moral circle beyond the magic number of 150. The near identity of the interests of social leadership with the interests of the gods met exactly the needs for social control brought about by the agricultural revolution, a trend that was only intensified with the emergence of still larger early city-states, kingdoms, and empires. Priests and kings replaced tribal units and charismatic shamans as the pressure for social control and regulation of human communities grew.

To trace the history of the ideas and rituals of sacrament and sacrifice in these religious developments, we need next to return to the cave paintings of Upper/Middle Paleolithic period. Most scholars believe these cave sanctuaries were the sites of seasonal cultic activities designed to endow the hunting and gathering communities with contact and control over food supplies. The mysterious forces governing herd animal migrations, the fertility of the lands, the abundance of prey animals, and the propagation of gatherable food stuffs were the targets of various rites. These rituals sought to establish and dramatize the right relationship between human beings and the land on which they depended. Supernatural forces were brought into symbolic communal relationship with people via the dances of the shaman and the people. A sacramental relationship was established through a "sympathetic causation" of ritual and actual hunt, the fertile land's providential bounty, and the propitiation of the force(s) that governed both. Since all psychological (e.g., mental and/or spiritual) processes can only occur in embodied form, the religious idea of community and communion of people and land must take physical form in dance, mask, animal hide, and offerings of meat, bone, first fruits, and blood. These tokens symbolically connect people and the forces of providential benevolence. Sacramentality builds upon the capacity of humans to participate symbolically in the very forces on which the health and well-being of the social organism embedded in a larger world depends. It makes little difference if the target of cultic attention is the hunted animal, the gathered tuber, or the harvested plant. The structure of the religious behavior remains the same.

Sacrifice is the inverse of this structure. Whereas sacrament is a positive participation of the community in divinely provisioned gifts, sacrifice is the symbolic repayment by the human community to the forces that governed the land's productivity. Sacrifice appears to have been carried out in a variety of ways. A partial list would certainly include immolation of a meat or grain offering, the spilling of blood by a prized animal, a human sacrifice, or the ritual letting of blood by the participants. Sacrifice expressed the repentance of the social unit for its necessary destruction of resources, especially as they feared evoking the displeasure of the supernatural agent responsible for the land's productivity. If sacrament was the expansive moment of life's unfolding, sacrifice was the setting-of-limits to how expansive the human community would become if it were to remain in right relationship with its world. It is in this sense that I refer to sacrament and sacrifice as the positive and negative feedback loops of religious community.

Christian Sacrament and Sacrifice

It is unlikely that any formal explanation of sacrament and sacrifice preceded the earliest Christian rituals of baptism or eucharist. The proto-church simply gathered for collective worship, and various rituals are likely to have evolved as they seemed appropriate to the kind of community these groups envisioned. Christian scriptures offer some evidence as to the emergent practices of the early church. St. Paul paints baptism in very dramatic terms, emphasizing that participants were to think of themselves as dying and rising with Christ. He emphasizes the threshold nature of the transformation, the leaving-behind of an old life for a new condition of ritual purity, and the importance of an inner conversion of spirit. Similarly, practice of the eucharist is portrayed as participation in the risen body of Christ and the ritual remembrance of the last meal enjoyed by Jesus with his disciples. Believers were reminded that Jesus was reported to have designated the cup as "a new covenant in my blood" (1 Cor. 11:24). These two rituals, along with a host of rapidly developing cultic practices, evolved organically from the needs of communal existence. They underscored the appearance of various novel qualities and gifts in the assembled believers as they undertook their religious journey. They were certainly important mechanisms to achieve and maintain social solidarity in the pluralistic Greco-Roman world in which Christian churches took root.

Christian understandings of sacrifice emerged in a similarly organic progression out of the matrix of Israelite ritual practices and beliefs, albeit with the novel twist that the Christian founder was conceived as both priest and self-offering. St. Paul encouraged an association between the Israelite rites of blood-offerings and Jesus' death on the cross (Rom. 3:25-26). Jesus' crucifixion was quickly contextualized as a self-offered death for "the remission of sins," an act that both entailed a loss of life and a reappropriation of that life in a transformed state that might be transmitted to all those who followed him in his risen body, conceived as the church itself. St. Paul emphasized the all-sufficiency of this singular and perfect blood offering and encouraged his converts to open themselves in fundamental trust to the One who purchased them from sin into mystical communion with the living Lord.

The Christian tradition's understanding of sacrament and sacrifice evolved against the background of ancient Israel's sacrificial systems and the cultic practices of the Greco-Roman world, especially, it seems, the so-called mystery cults. The word "*sacramentum*" originates in Roman law and designates a legal-religious sanction whereby an oath-taker would place his life and property into the hands of supernatural agents who policed fairness in exchanges and contracts.[8] In the early church it was associated with the Greek notion of *mysterion*, which was derived from the mystery cults and interpreted as an efficacious sacred sign or symbol. Its first attested usage in a Christian context occurs in the writings of Tertullian and is dated around 200 C.E.

Nothing like a formal theory of sacramentality appears in the Christian tradition until Augustine's *On Christian Doctrine* in the fifth century. He argues for a "visible sign of an invisible grace" interpretation, emphasizing the interior transformation of the human spirit through material representation in water, bread, and wine. Similarly, he explains the Christian doctrine of sacrifice in terms of the "true" sacrifice desired of God as "the love of God and neighbor." All later developments of Christian understandings appear to have built upon Augustine's seminal contributions.

An internal debate surrounding potential "magical" interpretations of the sacraments goes very deep in the Christian heritage. It was essential to the early church leaders to distinguish Christian practices from those of their competitors in the rich "pagan" world of the Roman Empire. Frequent commentaries attempt to draw a clear line of demarcation between the animal sacrifices of pagan rituals and the Christian practice of the eucharist. The principal difference was identified as the fact that no actual blood was used in the ritual and that the ritual's true meanings were entirely "spiritual." Nevertheless, the rituals were clearly portrayed as supernatural interventions in both contexts, a factor that persists to this day and raises serious questions for the modern Christian. Further, even with the frequent disclaimers, the Christian sacramental practices display a constant tendency to "decay" into magical or superstitious beliefs and practices, a charge powerfully raised by the early modern Enlightenment critics of religion.

The combined impact of the Renaissance, Protestant Reformations, and the Enlightenment in the modern period threw the idea of sacrament and sacrifice into disarray. Humanists such as Erasmus were openly contemptuous of magical and superstitious abuses in the church of his day. Protestant Reformers were similarly dismissive of the received doctrines of sacrament and sacrifice, although such criticisms were couched in their polemical engagement with Roman Catholicism. Enlightenment figures such as David Hume and Immanuel Kant picked up this line of argument in their writings on miracles and championed views of sacraments consistent with the growing power of the scientific age. For these writers, the sacraments became simple ritualized reminders of the calling of the Christian to live moral lives, an interpretation meant to underscore their reasonableness. By the time of the nineteenth century, post-Enlightenment Romantics such as Schleiermacher had recovered the distinction between external sign and internal intention/conversion in their interpretations. Further, there was a renewed emphasis on the communal nature of Christian sacramentality. Great fascination with the power of ritual and liturgy to shape human personality was to be found, especially in the English context, in the towering figure of John Henry Newman. These forces combined to bequeath a diverse and slightly confused understanding among the modern churches on the nature and role of sacrament and sacrifice in the current church.

Naturalistic Sacramentality

At the beginning of the twenty-first century, we enter a novel stage of interaction between religion and science, which demands, I think, bold rethinking of the Christian witness. Arthur Peacocke's Essay signals a willingness to rethink the meaning of Christianity in an intellectual climate where supernaturalism seems ever more suspicious. His Emergentist–Naturalist–Panentheism opens new pathways for understanding the ancient religious practices of sacrament and sacrifice. Above all else, it signals the need to understand Christian doctrines and rituals in a manner that eschews supernaturalism yet does justice to the experiential power of sacrament in religious community. In what follows, I will attempt to offer a naturalistic interpretation of sacramentality consonant with modern scientific theory, and especially with the current fascination with emergence.

When I recounted the story of sacrament and sacrifice in human evolution, my purpose was to resituate Christian ideology against the larger backdrop of the emergence and unfolding of life on our planet and the deep history of humanity among the interconnected species of our planet. This recontextualization expresses the first necessary step toward a consilience of religion and science in our contemporary intellectual climate. If we are to ask ourselves what occupies the most far-reaching narrative of humanity, we must grant that standing to the evolution story.[9] Nested within this story's branches we discover the primate lineage of which humanity is one offshoot. It is in this context that human cultural evolution appears as complex social existence and language usage molds the sapian mind/brain. The cultural twists and turns of this history mirror the convolutions of the human cerebrum itself. In the midst of the many cultural emergences grows Christianity, one among many blossomings of the human reaction to the process of negotiating a livelihood in the biosphere. In the light of this narrative, it becomes increasingly difficult to engage in any form of anthropocentric and triumphalist posturing before the complex network of relations that are our lifeworld. Similarly, nested among the beauty of many cultural flowerings, Christian identity can no longer craft its own story as an ascendant acquisition of cultural hegemony among other cultural expressions.

My presentation of the religious practices of sacrament and sacrifice in the history of our species told a story of cultivating and maintaining right relationship with sacred order or law. The principle underlying sacramental and sacrificial ritual is the mutual interdependence of human species and environment understood in a religiously valenced narrative. As historian of religions E. O. James notes:

> In this cycle of sacramental ideas and practices, as in their counterparts in the institution of sacrifice, the giving, conserving and promotion of life are fundamental. Thus, the eating of sacred food, like the ritual shedding of blood, is a means whereby life is bestowed through vitality inherent in it, and at the same time a bond of union is established with the sacred order.[10]

In his seminal book, *Religion Is Not about God*, Loyal Rue argues that all religions tell stories integrating a culture's ideas about reality and value:

> The narrative core is the most fundamental expression of wisdom in a cultural tradition—it tells us about the kind of world we live in, what sorts of things are real and unreal, where we came from, what our nature is, and how we fit into the larger scheme of things.[11]

Chaos is rendered as cosmos, a narrated and therefore meaningful world in which our understandings of who we are can be fused with what we must do in response to the nature of that world-situatedness. That story is best told in our day and age through our common evolutionary heritage as a species among species. Our presence is "miraculous" not because it is an exception to the way the world is, but *because* this is the way the world is, fecund with, and generative of, emergent properties. In Rue's artful phrasing, the evolution story tells us we are "star-born, earth-formed, fitness maximizing creatures endowed by natural selection with a set of species-typical traits for negotiating a livelihood on this planet."[12] To be Christian in the scientifically informed age is to see the Christian story as a chapter in this longer and larger narrative. In a word, it is about vision and the behaviors that result from a world well seen. Peacocke's *ENP* proposal is such a sacramental vision.

Sacrament and sacrifice are the ritual expressions of religious community when it sees its enmeshment in place understood as culture, environment, planet, and cosmos. Sacrament is the positive expression of giftedness, of a species in receivership of the blessings of good community, good land, good society, and good world. Sacrifice is the negative expression of limit, of a species that voluntarily, and for the well-being of community, land, society, and world, restricts itself to maintain right relationship with the sacred order. Like the feedback loops of any self-organizing and sustainable system, both moments operate in tandem, directing collective emergence and trimming excess. In our condition of environmental overshoot, perhaps no more salutary religious practice can be instituted than trimming our profligacy in voluntary sacrifice.

Just as important to an understanding of sacramentality in our context is its emphasis on the ordinary, properly seen, becoming the extraordinary. First, the physicality of water, bread, wine, and oil situates human spirituality in the midst of its carnal, embodied condition. In the biophysical experience of immersion in water, in communal sharing of bread and wine, in the sensation of warm oil on the scalp, we are thrown into our sensuality. In his marvelous book, *The Spell of the Senuous*, environmental activist David Abram speaks of reinhabiting our creatureliness:

> We have forgotten the poise that comes from living in storied relation and reciprocity with the myriad things, the myriad *beings*, that perceptually surround us . . . a common sensibility shared by persons who have . . . "fallen in love outward" with the world around them . . . A genuinely ecological approach . . . strives to enter, ever more deeply, into the sensorial present. . . . To return to our senses is to renew our bond with this wider life, to feel the soil beneath the pavement, to sense—even when indoors— the moon's gaze upon the roof.[13]

Second, the encounter with the ordinary foods of bread and wine, with the ubiquity of water, and with the warming touch of an oiled hand to the head, we are trained to see the simple realities of material presences in a new light. In our anti-supernatural age, we cannot viably conceive sacraments as magical invasions of a world otherwise devoid of divine presence, but rather we can think of the impact of the encounter in terms of retraining how we see. While this interpretation is likely different from much of the tradition's understanding of sacraments, it is entirely consistent with Augustine's early insistence that the sacrifice required by God is the human heart.

Sacrament and Science

During May and June of 2006, I conducted interviews with seven working scientists on my campus at Juniata College.[14] The interviews are the basis for a film project on which I am currently working called "Science and Sacrament." In the interviews we talk about the culture of working scientists, their relationship with their students, the intellectual and emotional satisfactions they derive from their work, and the aesthetics of scientific discovery. In this final section, I want to summarize some of what I have discovered from these discussions. My objective is modest: to display my hunch that the gap between religious sacramentality and the experiences of everyday scientific culture are by no means incorrigibly wide, and may in fact be capable of a novel forms of bridging and partnership.

In the interviews, I asked how these scientists had come into the world of their disciplines. One common theme in their responses was the joy of "geeking out" over their various areas of study. Jay Hosler, a biologist who specializes in honey bees and neurobiology, claimed scientists who make a difference model "a willingness to be a nerd." Ask them at a dinner party about some area they study, and great scientists light up and "talk like a kid." Students who can cross over the coolness barrier are the ones, in Hosler's experience, who "are going to find something that makes them happy for the rest of their lives." The other scientists I interviewed echoed this perception. They expressed it in terms such as "Boy Scouts" who collect minerals, who tinker with machinery for fun, or who ruminate over puzzles for hours, days, or years on end. What was most intriguing in the interviews were those moments when these scientists became immersed in describing their own areas of research. Without exception, it was here that the animation in their voices rose, their hands began to fidget, and their body language conveyed excitation. They modeled in the interviews what they admired in other scientists. They "geeked out" in the interviews and came alive in a manner that more abstract forms of questioning did not seem to provoke. In true "geek fashion," these scientists leapt to words like "cool," "really neat," and "awesome" to express how they felt when they were in the presence of really good science.

What I want to underscore with these observations is the element of vulnerability in these transformations. When Hosler says there is a barrier of coolness which must be overcome, he is referring to a capacity to let oneself go before an aspect of reality which for others is of trivial concern. If sacramentality is about encounter with an aspect of reality seen in the context of powerful life enrichment, if it is about vision and the blessings of relationship empowered by transformed vision, then there are close relationships between the vulnerability of the communicant and the geeking scientist. The high and the lowly are united in their awe before the vision of forces much greater than mere human conventionality.

One difference that emerged the interviews, however, was that the nature of the vulnerability was less morally tinged. In many religious sacramental moments, moral guilt comes before the perception of sacred forces. This is especially true when communicants assemble in the name of a supernatural agency before which believers are encouraged to think about their own unworthiness. In the Lutheran tradition of my heritage, for example, we began worship with the confession of sins: "I have sinned against you in thought, word, and deed, by what I have done and left undone. . . ." By contrast, the scientists I interviewed spoke of vulnerability not in moral terms but in social terms. It entailed getting worked up in unusual ways about things peers do not perceive as exciting and speaking intellectual jargon ("geek-speak") in contexts that were not appropriate. In fact, one physicist told the story of his break with the church in just these terms. As a child in the Episcopal Church, physicist Jim Borgardt was called forward and loaded with hymnals to make a point about his need for God. His contrarian response was to hold as many books as he possibly could. Only after the hymnals were stacked over his head, Jim narrates with pride, did they tumble down. The reaction of the priest was immediate: See, you need God, he roared. For Jim this was a decisive moment that he remembers vividly more than thirty years later. Further, it is closely associated in his mind with preachy dogmatism in church life and its lack of openness, especially to free thinkers. Jim celebrates science precisely for this embrace of the nerd and free-thinker and its lack of moralizing.

A different aspect of the same distinction is the nature of the challenge in the sacramental moment. All the scientists spoke of the great joy of the puzzle-dimension of their work. Many of them confessed to subscribing to special puzzle magazines beyond their work. Chemist David Reingold, for example, spoke about his joy in working on certain chemical puzzles in his discipline for time spans in excess of fifteen years. "I haven't really solved one yet," Reingold told me. Every puzzle in his area always breaks open novel questions, applications, and unforeseen challenges. The other scientists expressed similar delight and satisfaction in moments when a piece of reality disclosed itself to them through their circuitous puzzlings, when it gave up its mystery. There was no sense of the violation of the sacred as one might get from an invasion of the powers of the Tabernacle in a Catholic altar. For Borgardt, who has extensively studied the philosophy of science, this is the core difference between what he called Albert Einstein's vision of religion and science and his experiences in organized religions. For Einstein mystery invites exploration and there is a faith that humanity can think "the mind of God." The reverence scientists feel for their subjects comes without taboos, it seems, and is open to our play. For Borgardt and the others, there is no contradiction between the serious play of scientists with nature and the religious experiences of awe and reverence.

So what are the "sacraments of science," if we may use that expression? I discovered that they were as diverse as reality itself. What was most intriguing to me was that each scientist had a favorite phenomenon: the chemist had a favorite class of molecules (catenates); the geologist had a favorite mineral (copper); the physicist had a favorite experiment (double slit experiment) or physical display (lasers); the biologist had a favorite species (honey bees); the psychologist had a favorite experimental domain (running rats). Often these attachments derived from personal biography and were associated with pivotal moments of awakening in their own identities as scientists, what they universally called their "aha moments." These were their visible means to invisible realities, assuming that we take the visibility/invisibility distinction not as a metaphysical distinction but as a psychological one: these breakthroughs were the moments when a level of reality became visible to them of which they had not been mindful until that moment. This is, I contend, the heart of the sacrament in a scientific age. Sacramentality is about seeing relationships, not about the discovery of a supernatural agency at work in physical forms. The ordinary is seen as the extraordinary, the extraordinary does not elevate the ordinary.

But what of the sacrifice component? We have already discussed how the moral guilt aspect seems to drop out of the equation in a scientific paradigm. If sacrifice is not about expiation or atonement, how can we understand it? In my interviews, the closest association I found was with the cultivation of disciplined patience, the skills in the use of tools, and the rigorous logical training required to do good science. While creativity was certainly understood as important, so too were the skills of seeing well, working equipment well, mastering the intellectual theory that governed the research and the field, doing the complex mathematics describing the phenomena studied, and so on. Jay Hosler summarized this point in arguing that "it's not enough to be imaginative, you have to have the tools of the imagination" to do good science. In this sense, sacrifice was the retraining of the eyes and ears, the sharpening of the observational capacities of the scientist where sacrifice surfaces. Many, although not all, understood this as a preamble to a conversion to environmental care and concern. To love the world and its well-being, one must come to understand its beauty and intricacy in disciplined and patient exploration. Like the shaman who is undone ritually to be reborn as a seer, science requires a reassembly of the human imagination to see what science has to show.

Finally, it became clear that science generates its own kinds of community and discourse. Perhaps because all my interviewees work within a small liberal arts college context, they warmed to the subject most when talking about teaching and their relationships with their students. Chemist David Reingold was most animated, for example, when he spoke about an organization of five synthetic chemists he created, all of whom were in similar contexts of isolation. For several years, this group has assembled with their students to talk about their work in synthetic chemistry. Not only was it thrilling to gather with people of like interest, but it was also heady to be able to speak each other's language, to bounce problems off each other, to feel the sense of unity of purpose. Reingold ascribed to this group's existence the fact that many more of his students are now pursuing advanced degrees in synthesizing theoretically interesting molecules. Likewise, physicist Jamie White was most enthusiastic about his recent sabbatical work with a team of scientists working on novel laser technology in New Zealand. White said that he wants to give to his students a sense of what he had there. A similar emphasis on the importance of the team in the process of scientific discovery echoed throughout the other interviews. Perhaps it is stretching definitions too much to refer to these gatherings as the scientist's church, but certainly the camaraderie of co-participants, each performing his or her tasks toward some larger outcome, reminds one of St. Paul's body metaphors of functional diversity in the social organism.

The Common Fire

In closing, religion and science express the common desire, the common fire in the mind to find order in chaos. We may speculate that the evolution of both religion and science is rooted in the common evolutionary history of the emergence of mind to read a landscape and to see relationships between our species and the forces of the natural world on which each depends. In modern history religion and science have diverged, but the underlying impulse seems the same: to connect ourselves to networks of relationships that are greater than us. If religion is associated with a preoccupation with supernatural causes, and science has evolved to explain the world through natural causes, these two worlds cannot hope to meet. Arthur Peacocke's proposal of an Emergent Naturalistic Panentheism offers a hopeful model to rediscover the underlying causes and motivations of these two realms of human pursuit.

My concern in this brief chapter has been to promote an understanding of sacrament and sacrifice as naturalistic phenomena, as ritual expressions of religious community as it sees itself enmeshed in relations with the natural world. Sacrament redirects attention to the giftedness of humanity in its dependence on and enmeshment with its supporting environment, and sacrifice redirects our attention to the limits of human expansion. I also suggested that these cultural practices be understood as analogous to the self-assembly processes of complex, adaptive systems we find in nature. Finally, I tried to probe some similarities between the experiential world of religion and science with some ethnographic data from my interviews with scientists. I would not want to push any part of my suggestion in these final comparisons too far. They represent, at best, hunches about ways to explore novel understandings of sacrament and sacrifice in a scientific age where Emergent Naturalistic Panentheism is on the minds and hearts of religious seekers. If they provoke others to explore them further, then their purpose has been satisfied.

The Juxtaposition of Naturalistic and Christian Faith: Reappraising the Natural from within a Different Theological Lens

Ann Pederson

"Follow him through the Land of Unlikeness: You will see rare beasts and have unique adventures."

—W. H. Auden[1]

I have accepted Arthur Peacocke's invitation to join in the interpretation of his Essay and hope as well that new theological insights might be attained. And so I responded to Peacocke: "In engaging in this exercise, it has become clearer to me that, all along and presumably because of my *Bildung* as a Lutheran systematic theologian, I have been seeking a naturalistic formulation of the content of the Christian faith,[2] as far as can be achieved while doing justice to the data on which science rests. While I admit that I often have a desire to create a systematic whole, as is the hazard of my vocation, I would hope that my response to Arthur is open-ended and spiritually in tune with the spiritual explorations of many in our times."[3] The form of my response is similar to the musical form of sonata allegro in which the themes of the composition are stated in the exposition, then developed, and finally recapitulated. I see this composition as an interpretation of the themes exposited by Arthur Peacocke, whose own *opus* is clearly dedicated *soli deo gloria*.

Exposition

In the introduction to his Essay Peacocke hopes that his theological offering will resonate with the "spiritual explorations of many in our times."[4] Some thirty three years before, in 1972, Joseph Sittler, a Lutheran theologian, captured the longings, fears, and urgency of the spiritual times. And now his words seem even more appropriate: "No earlier time has had the knowledge or power to put its manipulative hand upon the dynamics of evolution or upon the molecular structures of matter and energy. But our time does, and a Christology that does not propose the power and presence and grace and judgment of God in Christ with an amplitude congruent with these power potentials as an operational mode of life deeply formative of technological man's personhood will be an unintelligible Christology, even an uninteresting one."[5] What we urgently need at the beginning of the twenty-first century is a Christological vision that can shape and inform a new and powerful way of helping humankind to interpret its place within the universe. A Christological vision that is unintelligible and uninteresting can have a profoundly deleterious soteriological implication: the orbit of God's saving grace will not be wide enough to encompass the universal place of humankind.

In an American culture marked by individualism, consumerism, and fear, some Christians have reduced the soteriological claims of the Christian faith to a "personal" need for a Lord and Savior. A reductionistic Christology omits the rest of the created order (and most others who are not Christian). Too many Christians are ready to abandon this world for a heavenly one, ignoring the dangers that we face living on this planet. A narrow vision, a narrow Christology. Much more is needed, however. Peacocke's vision reaches much further and deeper; it conceives a Christ who embraces the entire natural cosmos in all of its particularities. Where does he begin his Christological ruminations?

I once heard it said that some theologians begin with God (the doctrine of) and get to Jesus (Christology), while others begin with Jesus (Christology) and move to God (theology). Arthur Peacocke's move is clear and to the point: only when the foundations and universal scope of God's grace are fully established for all of creation, then and only then can the importance of God's specific work in Jesus the Christ be established.[6] The efficacy of God's work in Christ is first and foremost for *ta panta*—all of creation.[7] Peacocke formulates a theology of nature that is coherent and adequate, intelligible and interesting, both to the scientific world view and to fundamental claims of the Christian faith.

For Peacocke the natural world (the entire cosmos) is the scope and foundation for developing a Christology adequate to the spiritual longings of our times. Joining a host of other theological voices, he draws upon themes found in Eastern Orthodoxy to explicate his vision (cf. the chapter by Christopher Knight in this volume). The East and West emphasize different aspects of the work of Christ, the scope of God's grace, and the nature of the human person. Sittler explains these differences clearly:

> In the West that work is centered upon redemption from sin; in the East it is centered upon the divinization of man. In the West the doctrine

central to that work is atonement; in the East the central doctrine is participation, illumination, reenactment, and transformation. In the West the work of Christ is spoken of chiefly as restoration; in the East the work is reunification. The Western *Savior* is the Eastern *Pantocrator*.[8]

For many of the Christian theologians in the religion and science dialogue, the vision of the Eastern Fathers seems the most theologically adequate for relating to science, which describes the world as dynamic, evolutionary, and emergent.

Peacocke begins with St. Irenaeus's theological category of recapitulation (*anakephalaiosis*), and his themes of *imago and similitude,* to formulate his own naturalistic Christology. Indeed Peacocke might agree with Sittler that, "This is not a 'natural theology' in the sense that God is disclosed in nature without the revelation in the Son; but it is a *theology for* nature in the inevitable sense that the hand of God the creator, which is the hand of the Son, should be seen, following the incarnation, also in nature."[9] The purpose of God's grace for all of creation comes to fruition in the incarnation. Nature is a means of grace. Irenaeus developed his understanding of the universal scope of God's grace from within his arguments with Gnosticism (which still rears its head under different guises in our own culture).

Irenaeus argued with Gnostics, who viewed the body and nature as a source of evil. To counter their view, he reemphasized the goodness of creation as a central theme in his theology. Jesus the Christ, who is the incarnation of God in the natural world, creates, blesses, and redeems creation. Rosemary Ruether, a feminist-liberationist theologian, makes this clear:

> For Irenaeus, creation is itself an incarnation of the Word and Spirit of God, as the ontological ground of bodily existence. The incarnation of the historical Christ is the renewal of this divine power underlying creation. In the incarnation divine power permeates bodily nature in a yet deeper way, so that the bodily becomes the sacramental bearer of the divine, and the divine deifies the body.[10]

The body of Christ is not just human flesh; it includes the flesh and blood of all creation. All things (*ta panta*) are created and redeemed through God's sacramental and incarnational power. The body of Christ is personal and communal, human and nonhuman, natural and technological. God's creating and redeeming grace permeates all creation deeply and God calls it good.

Peacocke also develops the Eastern Orthodox notion of *theosis* to explain the relationship between God and the world. Veli-Matti Karkkainen, a Finnish theologian, cites two Patristic texts that form the theological foundation of *theosis.*

With regard to *theosis*, the two patristic texts most often cited are from Irenaeus and Athanasius. Irenaeus: "The word of God, our Lord Jesus Christ . . . did through his transcendent Love, become what we are, that He might bring us to be even what He is Himself." Athanasius: "he, indeed, assumed humanity that we might become God."[11]

All of creation participates in the divine life, which is an intercommunion of multiple species. The interesting and rather radical notion of Peacocke's development of *theosis* is that what is natural is not only fully part of the creation but also fully part of the divine. God becomes fully human/natural in order that we can become fully divine/natural.

Peacocke utilizes *theosis* to *naturalize* the sacramental and incarnational relationship between God and the world. We are not resident aliens in this world of God but fully naturalized citizens in our natural home. The quote from Traherne that Peacocke uses helps to explain this idea further: that the "WORLD is unknown, till the Value and Glory of it is seen: till the Beauty and Serviceableness of its parts is considered."[12] When one enters the world, one loses one's self, only to find God in exchange. This is a powerfully poetic and mystical manner of interpreting *theosis as a fully natural and worldly process*.

Therefore, both the person and work of Jesus the Christ must be *natural* in order to redeem that which is natural. Throughout the history of the Christian tradition, the church has argued over what it means for Jesus the Christ to be both fully human and fully divine, but I have found very little evidence to indicate that nature itself has been part of the discussion. Peacocke, however, reclaims nature as the matrix for understanding both human and divine.[13] This provides creative ways for reformulating the divine/human relationship and the incarnation of God in Jesus the Christ. God's incarnation in the natural world should not be opposed to or radically different from God's incarnation in the person of Jesus the Christ. In fact, as divine and human participate in the process we are calling *theosis*, it might be that both divine and human are so transformed and transmuted that the categories themselves become quite transformed.

The writings of Martin Luther and their subsequent elaboration in the contemporary dialogue between Orthodoxy and Lutheranism are also a helpful resource for interpreting *theosis*. The image of *theosis* corresponds to the Lutheran Reformation language of *finitum capax infiniti*. Luther affirms that the finite is the bearer of the infinite against those who maintain the contrary (and also against Aristotle). In his "*Die Disputation de divinitate et humanitate Christi*," Luther argues against those who assert that there is no "proportion" between the created and the Creator, the finite and the infinite. Luther claims there is indeed unity: "*Non stamen non tantum facimus proportionem, sed unitatem finiti et infiniti*."[14] The natural world indeed bears the infinite company of God. The Word becomes flesh—this is where we know for sure that God will be present.

Closely related to the language of *finitum capax infiniti* is Luther's language of the "blessed exchange." This occurs when God in Christ takes on our humanity in order that we can take on God's divinity, albeit in a cruciform shape. The blessed exchange transforms us into "christs," divine incarnations of God whose purpose is to serve the neighbor. Peacocke, along with other theologians like Philip Hefner, describe humans as created co-creators. If the blessed exchange or the image of *theosis* leads to this anthropological extrapolation, might it not also lead us to say that humans are also redeemed co-redeemers? As human beings, we are called to become partakers and participants with God in the ongoing creating and redeeming work of God. At the end of his Essay Peacocke agrees that this transforming grace of God is kenotic, cruciform: humans take upon themselves Christ, but always in a cruciform shape. To be human is to be divine but not in ways that we expect.

God's incarnation in the natural world often takes a cruciform shape, suffering with and transforming the pains of creation. Peacocke claims that we are called as priests to serve the creation from whatever specific place we are located. The sacramental words of grace, "given for you," are not just for humans but are blessings of grace for the whole of creation. The "giftedness" of the world from God becomes our offering back to God and to the world:

> All of the foregoing, any consideration of the operation of transformative, divine grace in the world in general, and in human beings in particular, has to be placed in this kenotic context—the recognition that divine grace is costly to God and so involves a sacrificial element which has been variously expressed liturgically, theologically and aesthetically.[15]

God's presence in the world is radically immanent and radically natural (which is why it is also radically transcendent). When the divine embodies all of creation, then all of creation, in all its beauty and brokenness, is redeemed. And this redemption is not separate from God's intentions established in creation. Creation and redemption are processes of participation of the natural communities in which we live, move, and have our being. The layers and webs of relationships in creation are fully natural in both their beauty and brokenness. When we enter into the joy and sorrow of the natural world and lose ourselves, we find God present in us through the power of transforming grace. We emerge again, divinized, naturally.

Panentheism corresponds to this view of God's sacramental purposes in the world. The divine is transformed into an immanent shape of the natural world:

For the panentheist, who sees God working in, with, and under natural processes, this unique result (to date) of the evolutionary process corroborates that God is using that process as an instrument of God's purposes and as a symbol of the divine nature, that is, as the means of conveying insight into these purposes. But in the Christian tradition, this is precisely what its sacraments do. They are valued for what God is affecting instrumentally and for what God is conveying symbolically through them.[16]

Sacraments are by their very nature enacted in embodied communities. We become co-creators and co-redeemers with God in the business at hand. The sacraments are a sign of God's naturally creative action in the world.

Development

To tell the story of the natural world is to weave a narrative that is complex and tangled with threads of culture and technology. As Peacocke notes, the narratives from the scientific world tell us that we have emerged from these tangled webs whose boundaries between human and the rest of the natural world are blurred. When we tell this story of the natural world and of the human place within it, we are required to jump into the middle of the story because the beginnings and endings are mixed together. To consider who we are as human beings requires a new kind of narrative imagination.

Peacocke's panentheistic view of God and the world challenges the traditional Western classical theistic view:

Within the framework of an emergentist–monist–panentheistic–naturalistic (*ENP*) perspective, one no longer needs to choose *between* categories such as God, the world, matter, energy, and information. Instead, one can hold them all together in a new kind of synthesis that obviates many of the false dichotomies—the sciences versus the humanities, matter versus the spiritual, science versus religion—which have plagued Western culture for too long.[17]

While I find this metaphysical world view quite convincing, I believe that the particular implications of it need more explication. To accomplish this task I will join the ideas of Donna Haraway, a feminist philosopher of science, with Peacocke's. Donna Haraway might find it odd that her work would be used by theologians (since she has rejected theism) to develop and expand the idea of what is "natural" and what is "created." And yet her upbringing as a Roman Catholic and her constant references to incarnation and embodiment suggest that her categories are indeed very consonant with what we are up to in this volume. I'd like to use Haraway's work to help develop the notion of what is meant by "natural" so that the soteriological implications of Peacocke's work have a deeper and more particular scope in which Christ is present.

Donna Haraway, a professor of history of consciousness and feminist studies at the University of California, Santa Cruz, envisions new ways of thinking about the world in which we live. As a cultural critic and philosopher of science, she develops metaphors that help us explore relationships between nature, culture, human, animal, and machine. In her most recent writing about the nature of humans in the natural world, she has moved from her use of the metaphor of cyborg to the relationship between human and dog as co-companion species. Her images blur traditional Western understandings and boundaries, most often expressed in binary dualisms. She draws upon her love of Alfred North Whitehead and feminist thought to develop her position:

> My love of Whitehead is rooted in biology, but even more in the practice of feminist theory as I have experienced it. This feminist theory, in its refusal of typological thinking, binary dualisms, and both relativisms and universalisms of many flavors, contributes a rich array of approaches to emergence, process, historicity, difference, specificity, co-habitation, co-constitution, and contingence.[18]

Haraway's highly creative work can aid us in deriving further theological insight from Peacocke's descriptions and implications of the natural world. One can see shared intuitions and interests in their visions.

If we think that the familiar world of the Enlightenment is collapsing for the West, then we might find ourselves sympathetic to Haraway's insights. For Haraway the movement from the Enlightenment world view to a post-Enlightenment vision is marked by powerful implosions of traditional philosophical categories:

> If belief in the stable separation of subjects and objects in the experimental way of life was one of the defining stigmata of modernity, the implosion of subjects and objects in the entities populating the world at the end of the Second Millennium—and the broad recognition of this implosion in both technical and popular cultures—are stigmata of another historical configuration.[19]

Nature and culture, science and religion, human and nonhuman, subject and object—categories once thought separate—now implode in our post-twentieth-century historical, cultural, and political settings. Haraway's metaphors of primate, cyborgy, and companion species (dog–human) traverse and trick the traditional Enlightenment categories into new configurations of what is natural and what is created or made. In a similar vein, Peacocke's vision of an "emergentist "monist"–panentheistic–naturalistic (ENP) perspective" dissolves the rupture between categories and instead places them "into a new kind of synthesis."[20]

Pure categories don't exist (and really never did). Haraway writes:

> Cyborgs and companion species each bring together the human and non-human, the organic and technological, carbon and silicon, freedom and structure, history and myth, the rich and the poor, the state and the subject, diversity and depletion, modernity and postmodernity, and nature and culture in unexpected ways. Besides, neither a cyborg nor a companion species animal pleases the pure of heart who long for better protected species boundaries and sterilization of category deviants.[21]

Nature is *naturally* deviant, never pure-bred, and always familiar in totally unexpected ways. This naturally deviant, mongrel world of the technological, social, political, religious, and historical is the nature into which God's incarnation takes place and into which humans live, move, and have their being.

At the heart of Haraway's project is her critique of the way that American culture embodies and practices the commodification and objectification of individuals. She worries about "its mongrel technologies of purebred subject- and object-making."[22] Keeping categories and species neat and pure takes a lot of work and money. And yet she finds as she cleans up her dog's poop in the morning with her plastic wrapping from the *New York Times* morning paper that "pooper scoopers [are] quite a joke, one that lands me back in the histories of the incarnation, political economy, technoscience, and biology."[23] Like the ecclesiastical sorts who define human and divine by keeping them in neat compartments, away from the political, biological, and cultural messes of every day life, Haraway finds that our world is full of reminders that life is much more scatological then we want to think. Incarnation is about what is *natural* in all of its political, social, economic, religious, cultural, biological, scatological, and historical messiness—this is the finite which bears the infinite. As a former Roman Catholic, she notes ironically that nowhere else did the categories of human and divine, created and creator, implode more than in "living the relationship and speaking the verb that passes as a noun: companion species. Is this what John meant when he said, 'The Word was made flesh'?"[24] The Word and flesh are co-companion species, becoming and passing within each other as a living relationship of imploding grace.

Incarnation is a flesh made word and words made flesh. We become kinfolk with all of the created order in its muddled complexity. Haraway explains:

> At the origin of things, life is constituted and connected by recursive, repeating systems of information. . . . these flows, not the blood ties connecting bodies in another regime of nature, are the circulatory systems that constitute kinship—replete with all of its trans-hybridities and reworkings of race, species, family, nation, individual, corporation, and gender—at the end of the second Christian millennium.[25]

We are by nature mutts, hybrids, mongrels. When we talk about God becoming incarnate, taking on the flesh of the natural world, this includes all of these repeating systems of information—circulatory webs of flesh and blood co-mingled. "What cannot be assumed cannot be healed." To redeem that which is natural means that God in-corp-orates all those circulatory systems of kinship.

We become who we are through relationships over time and our becoming is a kind of ontological sacramental practice; we are always becoming in, with, and under. As technologically formed beings, our natural way of being in the world is in, with, and under the worlds of technoscience, "worlds whose fibers infiltrate deep and wide throughout the tissues of the planet, including the flesh of our personal bodies."[26] Human creation and redemption is linked in, with, and under the Word whose fibers move us deeper into the tissues and flesh of the planet (indeed of the cosmos).

Recapitulation

So what are the implications of this soteriological vision of the natural world posited and developed by Peacocke? For many people, naturalistic Christian faith would appear as an oxymoron—words paired together that cancel each other's meaning. For example, think about when someone is very ill and we say we must "let nature take its course." Or when we speak about the value of "natural childbirth" or "natural death"? Like the early docetists, we leave the flesh out of nature, out of both divine and human nature. We forget the dynamic webs of social, cultural, political, and technological relationships in which we understand human personhood. We prefer to think that our origins come from some pristine, pure garden of original nature to which we longingly wait to return. And yet we are placed "east of Eden" where the purebred, pristine boundaries of perfection are left only to idyllic memories. We are here—in this time and this place—in a world created by a God whose own incarnation within it is messy, bodily, and finite. We bear the body of God within our selves. We enter into the body of Christ given for us. We become living sacraments to one another. When we delight in God and in God's world, we find that the delight of self and other is given back to us as the sacrificial love of God. This kind of *kenosis* or *theosis* is the very opposite of what we often expect the divine to be. This divine discontinuity, however, is continuous with God's creative purposes.

Sacraments are naturally that which signifies how the finite bears the infinite: the divine, transformative grace of God working in the world. Peacocke writes:

> For those who aspire to co-create *with* God in nature—working harmoniously with the grain, as it were, of the natural order—will need more and better-informed science and technology in order simultaneously to provide for human needs and to respect the rest of the world, living and non-living. This, it seems to me, is the direction which we should be looking for a sound basis for that "creation-centered" spirituality and a theo-centric ecological ethic which many in our contemporary society are now seeking.[27]

To be christologically formed is to serve the neighbor—the natural world in all of its intricacies. In an odd way, this service for others becomes a kind of self-service:

> We would also have to be recipients of that divine grace that can transform individuals into creative community, to really becoming "church" in its most basic sense as the channel of transformative grace to all of the world, including those human beings not consciously members of it.[28]

I wonder if church is really ready for the radical *nature* of this vision. Mission is so much simpler when its scope is reduced!

God and humans are partnered together in an intimate relationship of serving the least of the world. Parker Palmer, an American Quaker spiritual author and educator, explains that the word *truth* comes from the word, "troth," as in, "I pledge thee my troth." To know the truth is to be pledged in "troth" to one another, to be fully engaged with another. How we know one another is how we know the truth:

> In truthful knowing we neither infuse the world with our subjectivity (as premodern knowing did) nor hold it at arm's length manipulating it to our needs (as is the modern style). In truthful knowing the knower becomes co-participant in a community of faithful relationships with other persons and creatures and things, with whatever our knowledge makes known. We find truth by pledging our troth, and knowing becomes a reunion of separated things whose primary bond is not of logic but of love.[29]

As a community called church, Christians are called into faithful relationships with all of creation, as co-participants with the Creator. I have a feeling that our liturgy, adult forums, Sunday School, and other gatherings would be radically different if we took our calling seriously to be be-trothed with all of creation, and not just among ourselves. Think then how mission might be shaped! Is the church ready not only to know this kind of radical gospel but also to practice it?

We have begun to address Sittler's opening question. Peacocke's Christological vision is interesting, intelligible, and inclusive of all life precisely because of its naturalistic scope. The question is whether or not the Christian community can embrace what God embraces—that we as humans are the naturalized citizens of the created order fully embodied, enfleshed, and *au natural.*

That which is natural is created and loved by God. And yet, as we have noted, what is *natural* is neither pristine nor pure. But neither is the Christian faith. Some Christians have wanted to find easy, pure, clear answers to the messy, difficult, impure dilemmas of our world. But such is not the case. For the divine answer to the world's questions comes *naturally* in the Word made Flesh.

Arthur Peacocke: Postmodern Prophet

Nancey Murphy

My hope in writing this essay was to intrigue Arthur enough that he would overlook the fact that I did not engage him at the theological level, which is his main contribution in his Essay. I have bigger fish to fry.

Escaping the Third Axis

I have spilt quite a lot of ink over the past years peddling a clever scheme for distinguishing modern from postmodern thinkers (in the Anglo-American world, at least). My late husband, James McClendon, and I decided upon three defining philosophical features of modern thought: foundationalism in epistemology, referentialism/ representationalism in philosophy of language, and individualism in ethics, political theory, etc. The clever aspect was Jim's recognition that these were assumptions that provided the bases for the major *arguments* in modern scholarship: between optimistic foundationalists such as John Locke and pessimists such as David Hume; between those in the theory of language who granted some significance (such as self-expression) to fields of discourse such as ethics and religion and those who took only factual language to be meaningful; and between individualists and collectivists with regard to political, ethical, or moral agency. This created three axes upon which any given thinker could be located, and these three axes could be combined to create a "Cartesian coordinate system" for locating modern thinkers. Anyone who got out of that space altogether was "postmodern."

We called the first two axes the epistemological and linguistic axes, respectively, but never had a good name for the third. I finally settled on calling it a metaphysical axis after studying Arthur Peacocke's work. I came to recognize that individualism in the human sphere is but an instance of the general assumption of atomist-reductionism (whose modern contrary might be some romantic valuation of the whole).

The postmodern alternative to foundationalism is epistemological holism, as in Thomas Kuhn's and Imre Lakatos's scientific methodologies. The postmodern alternative to referentialism is an understanding of the meaning of language as deriving from its use (on my view, critical realism is still in the "Cartesian" space). But what is the postmodern alternative to atomist-reductionism? Peacocke pointed the direction at least as early as 1979.[1] His work on top-down causation and/or whole-part constraint has since then been amplified and accepted by many.

Yet not everyone accepts the arguments. This is a worldview issue; it is in Ludwig Wittgenstein's terms a "picture" or "metaphor" by which we judge arguments. In this case it is the picture of a hierarchy of more and more complex arrangements of atoms. Must it not be the case that the movements of the parts determine the movements of the whole? It is interesting that in the relatively short time I have been teaching on these topics, I find that students' assumptions as they enter class are changing. Whereas I used to struggle to explain how reductionism (e.g., the explanation of phenomena in terms of fundamental laws and particles) could *fail* to be true, I now have to help half of the class see how anyone could have believed in it. If this is indeed a period of change from modern to postmodern thinking, the observable shift in assumptions regarding this issue adds strength to my claim that reductionism has been an important part of the modern worldview.

The Return of God

There are other accounts of the defining features of modernity, with correlative judgments about what counts as transcending it. I enjoy the irony in Bruno Latour's book, *We Have Never Been Modern.*[2] The first modern move, he says, was to distinguish nature from culture. Nature is transcendent, always "out there" to be discovered, not created. Culture or society, however, is what we freely make; thus we know it immanently. This distinction is too familiar to require much comment: it is the opposition of the objective versus the subjective, the natural sciences versus the social sciences, facts versus values. The double irony is that the laws of nature are only known as they are *fabricated* in the laboratory by the new *social order* of scientists, while society turns out to transcend the humans who created it—it has an *objectivity* of its own.

An equally important modern move was "the attempt to free intellectual pursuits from the influence of religion."[3] Latour claims that God had to be removed from "the dual social and natural construction." "No one is truly modern who does not agree to keep God from interfering with Natural Law as well as with the laws of the Republic."[4] This "crossed-out God" is distanced from both Nature and Society, yet kept presentable and usable nonetheless. Latour writes:

> Spirituality was reinvented: the all-powerful God could descend into men's heart of hearts without intervening in any way in their external affairs. A wholly individual and wholly spiritual religion made it possible to criticize both the ascendancy of science and that of society, without needing to bring God into either. The moderns could now be both secular and pious at the same time.[5]
>
> Latour's irony again.

To recognize such a state of affairs, to date its historical beginning, is to begin to transcend it. But it is only a beginning. It calls into question the compartmentalization of reality—God, Nature, Society—but it does not tell us how to reunite them. It does not tell us how to proceed toward a more holistic view of enquiry, relating theology to both the natural and the human sciences.

This brings me to a second crucial and far-sighted contribution of Peacocke's work—the claim that theology belongs at the top of the hierarchy of the sciences. So far as I know, the notion of theology as the top science in a (thoroughly) nonreducible hierarchy originated implicitly in Ian Barbour's and explicitly in Peacocke's writings. It is implicit in Barbour's claim in *Issues in Science and Religion* that "an interpretation of levels can contribute to a *view of man* which takes both the scientific and the biblical understanding into account."[6] Here Barbour implies that the religious perspective is an indispensable level of description of human life. This notion was explicit in Peacocke's work by the time he published *Creation and the World of Science* in 1979. In an appendix he writes:

> It seems to me that no higher level of integration in the hierarchy of natural systems could be envisaged than [worship and other religious activities], and theology is about the conceptual schemes and theories that articulate the content of this activity. Theology therefore refers to the most integrating level we know in the hierarchy of natural relationships of systems and so it should not be surprising if the theories and concepts which are developed to explicate the nature of this activity . . . are uniquely specific to and characteristic of this level. . . . For this reason theories and concepts which the theologian may apply objectively to religion . . . have a right not to be prematurely reduced, without very careful proof, to the theories and concepts of other disciplines appropriate to the component units (society, man, nature, etc.), the unique integration of which in a total whole comprise the religious activity *par excellence*.[7]

The full spelling-out of theology's own proper content in relation to the disciplines below it is the primary focus of Arthur Peacocke's lead Essay in this volume.

The Autonomy of Morals

I turn now to a third feature of modern thought. This is the attempt to sever moral reasoning from the traditions upon which it depended in the past—sometimes from philosophical traditions such as Aristotle's, but most often from theological traditions. Stephen Toulmin and Jeffrey Stout, in complementary ways, explain why this appeared to be so necessary at the dawn of modernity.[8] If morals are dependent upon the authorities who speak for the religious traditions, then the result of the fragmentation of Christianity after the Reformation could only be the fragmentation of society, seen at its worst in the Thirty Year's War. The great hope was that the kind of universal human reason employed in the sciences would provide universally acceptable conclusions about morals and politics as well.

We see, at the end of modernity, that this was a vain hope. The rational systems associated with the names of Immanuel Kant, Jeremy Bentham, and John Locke provide conflicting guidance, and there is no rational way to choose one system over the others. Alasdair MacIntyre argues that the development of theories in philosophical ethics from Hobbes, at the beginning, to the Bloomsbury group in the early twentieth century, has been a failed attempt to provide a theoretical rationale for traditional morality. He concludes that modern moral discourse is in a grave state of disorder. He makes a pointed analogy: contemporary moral discourse is comparable to a *simulacrum* of science after a know-nothing regime has killed the scientists, burned the books, and trashed the laboratories. Later, fragments of scientific texts are read and memorized, but there is no longer any recognition of the *point* of science.

Similarly, MacIntyre says, our moral language is a hold-over from the past, but we have forgotten the original *point* of morality. In particular we have forgotten the context that once gave it its meaning. What we moderns (and postmoderns) have lost is any notion of the ultimate purpose or *telos* of human life. Such accounts of the human *telos* used to be provided by traditions, usually religious traditions. MacIntyre argues that the correct form of ethical claims is something like the following, *conditional* statement: "If you are to achieve your *telos*, then you ought to do *x*." It is a peculiar feature of modern Enlightenment views of ethics that their proper form has been taken to be apodictic: simply, "you ought to do *x*." Modern philosophers have developed competing theories regarding the most basic moral claims: "you ought to act so as to achieve the greatest good for the greatest number" versus "you ought to act so that the maxim of your action can be willed universally." But because morality is taken to be autonomous—that is, unrelated to other knowledge—there is no way to arbitrate between these most basic construals of the moral "ought." This impossibility results in the interminability of moral debates in our society. However, the interminability should not, says MacIntyre, be taken as the intrinsic nature of moral discourse, but ought rather to be seen as a sign that the entire Enlightenment project has taken a wrong turn. The wrong turn was the attempt to free morality and ethical reasoning from religious tradition. For it is traditions that provide the starting point for settling moral disputes. They provide the resources for answering the question: What is the greatest good for humankind? Is it happiness? Is it living in accord with the dictates of reason? Is it a just heavenly reward? Or is it more complex than any of these?

So theology or metaphysics provides a concept of the purpose for human life. Ethics is the discipline that works out answers to the question: how ought we to live in order to achieve our highest ends? In addition, MacIntyre argues, such theories of human flourishing can only be fully understood insofar as we know how they have been or could be socially embodied, so the *social sciences* are the descriptive side of a coin whose reverse, normative, side is ethics.[9]

Thus, MacIntyre's contribution is to argue that the modern view that insulates moral reasoning from knowledge of the nature of reality, *both theological and scientific,* is an aberration. Ethics needs theology (or some substitute for theology), and the sciences, particularly the social sciences, need ethics.

Peacocke and MacIntyre in Conversation

A combination of insights from MacIntyre and Peacocke has led George Ellis and me to make two modifications in Peacocke's model of the hierarchy of the sciences. The hierarchy of the sciences is ambiguous as to whether higher levels pertain to more *encompassing* wholes or to more *complex* systems. These two criteria usually overlap, but not in every case. If the hierarchy (apart from theology) is taken to be based on more encompassing wholes, then cosmology is the highest possible level. If the hierarchy is based instead on increasing complexity of the systems studied, then we have to ask whether a social system or the human nervous system is not more complex than the abstract account of the cosmos provided by cosmologists. There is no good way to choose between these criteria; therefore it is helpful to represent the relations among the sciences by means of a branching hierarchy, with the human sciences forming one branch and the natural sciences above biology forming the other.

Our next move was to employ the concept of "boundary questions" to argue that ethics belongs in the hierarchy of the sciences, above the social sciences and below theology. A boundary question is one that arises at one level of the hierarchy but can only be answered by moving to a higher level. There are many that arise from cosmology, such as why there is a universe at all.

There has been a long-standing debate as to whether the social sciences are value-free. The debate goes back to the very beginning in that August Comte, who coined the term "sociology," understood the discipline to be the science of the improvement of society. Thus, a concept of what is good for society was an essential aspect of the new science.[10] Max Weber argued, in contrast, that blurring the distinction between fact and value led to prejudice.[11] Weber's view predominated through much of the history of the social sciences in the English-speaking world.

However, I believe that Charles Taylor has dealt a mortal blow to the concept of value-neutrality. In brief, Taylor argues that the human sciences study humans, and a distinctive aspect of human selves is that they engage in what he calls "strong evaluation." That is, they make judgments about right and wrong, better and worse, higher and lower, which are not rendered valid by their own desires or choices but are intended to stand independent of these desires and choices and offer standards by which they can be judged.[12] Consequently, the attempt at value-free description of human affairs is bound to fail; a self who can only be understood against the background of distinctions of worth cannot be captured by a scientific language that essentially aspires to neutrality.[13]

To illustrate the involvement of moral evaluations in the social sciences, Ellis and I developed several examples.[14] Consider the role of violence in sociology and political theory. The assumption that violent coercion is necessary to maintain society goes back at least to Thomas Hobbes's claim that the state of nature, prior to the social contract, is the war of each against all. A variety of theorists since Hobbes have maintained that coercion is necessary to preserve society, and that violence is merely the ultimate form of coercion. As Peter Berger says, "Violence is the ultimate foundation of any political order."[15]

Now, in what sense is this an ethical assumption? Is it not simply a statement of empirical fact, a law of human behavior? Reinhold Niebuhr, known primarily as a Christian *ethicist,* concurs with the majority on the impossibility of noncoercive, nonviolent social structures. His view is dependent upon a prior ethical judgment, a judgment regarding the highest good for humankind. This view of the human good is in turn the consequence of a particular theological doctrine. Niebuhr writes:

Justice rather than unselfishness [is society's] highest moral ideal. . . . This realistic social ethic needs to be contrasted with the ethics of religious idealism. . . . Society must strive for justice even if it is forced to use means, such as self-assertion, resistance, coercion and perhaps resentment, which cannot gain the moral sanction of the most sensitive moral spirit. . . .[16]

Niebuhr's judgment that justice is the highest good that can reasonably be expected to be attained in human history is in turn based upon his eschatology—his theological vision for the end of history. Salvation, the kingdom of God, the eschaton, are essentially *beyond* history. The reason Niebuhr takes this stand on eschatology, in contrast to a view of the kingdom as realizable within history, is that he has set up the question in terms of the problem of the temporal and the eternal. Since it is not possible to conceive of the eternal being realized in the temporal, he concludes that the kingdom of God is beyond history, and this, in turn, means that guilt and moral ambiguity must be permanent features of the interim.

So, contrary to claims for the value-free character of the social sciences, it takes but a little scratching to find ethical judgments under the surface, which cannot be evaluated by any scientific means. This is, instead, the proper subject matter of ethics.

Niebuhr's argument has nicely illustrated MacIntyre's claim that moral reasoning is essentially dependent upon some concept of the ultimate purpose of human life, and that such concepts come from theological traditions or from some substitute for theology. Marxism has a *telos,* a secularized kingdom of God. Scientific materialism has an account of ultimate reality. In Carl Sagan's memorable terms, the universe is all that is and all that was and all that ever will be. The connections between a straightforwardly theological account of ultimate reality and prescriptions for the good life are usually clearly drawn. In my Catholic days we memorized the *Baltimore Catechism's* answer: the purpose of life is to know, love, and serve God in this life and to be happy with him in the next. The connections between an account of the material universe as ultimate reality and prescriptions for living are more tenuous, but one version is that the best we can achieve is the courage to face the fact that the human race is a cosmic accident, and we must therefore *create* as much meaning during our short individual and species' lifespan as possible.[17]

Theistic versus Atheistic Naturalism

An interesting phenomenon in the public sphere these days is what one might call atheistic theologizing. Tom Ross pointed out long ago that Carl Sagan's work could be perceived as a naturalistic religion. He begins with biology and cosmology, but then uses concepts drawn from science to fill in what are essentially religious categories—categories, by the way, that fall into a pattern surprisingly isomorphic with the Christian conceptual scheme. He has a concept of ultimate reality: "The Universe is all that is or ever was or ever will be." He has an account of ultimate origins: Evolution with a capital *E.* He has an account of the origin of sin: the primitive reptilian structure in the brain, which is responsible for territoriality, sex drive, and aggression. His account of salvation is gnostic in character—that is, it assumes that salvation comes from knowledge. The knowledge in question is scientific knowledge, perhaps advanced by contact with extra-terrestrial life forms who are more advanced than we.[18] Sagan's account of ethics is based on the worry that the human race will destroy itself. So the *telos* of human life is simply survival. Morality consists in overcoming our tendencies to see others as outsiders; knowledge of our intrinsic relatedness as natural beings (we are all made of the same star dust) can overcome our reptilian characteristics.

Peacocke's hierarchical model explains this phenomenon. If God is removed from one's worldview, this does not remove the *space* at the top of the hierarchy waiting to be filled by something. For the typical naturalist, it is the universe itself.

This is an important insight. The atheist sees that he and Peacocke substantially agree about the character of the natural world, but he then asks what justification there is for *adding* God to it. If we recognize that every worldview has some implicit or explicit account of ultimate reality, then that account is equally in need of justification.

It happens that I was asked to respond to a paper arguing for atheism by Richard Dawkins. I imagined myself in the long line of apologists attempting to be the first to provide in one hour a knock-down argument for the existence of God, but knew too much history to entertain the thought for long. I decided instead to use the resources provided by Peacocke for understanding the relations between theology (or a-theology) and the sciences in order to "place" Dawkins in the intellectual world of our day. My starting point was the one just made above: if one takes the hierarchy of the sciences, as Peacocke describes it, and simply removes God, this does not remove the "space" where theology used to be. It calls for a replacement position on the nature of ultimate reality. And such a replacement needs to be argued for, just as does a theistic position.

James Turner makes a startling claim in his excellent book, *Without God, Without Creed: The Origins of Unbelief in America.* He argues that disbelief was not a live option in Europe and the U.S. until roughly between 1865 and 1890.[19] Philosopher Merold Westphal helpfully distinguishes two sorts of atheism. One he calls evidential atheism, well represented by Bertrand Russell's account of what he'd say if he were to meet God and God asked why he had not been a believer: Not enough evidence, God! Not enough evidence![20] Given the difficulties in adapting theological reasoning to modern canons of rationality, this response is readily understandable.

But if religious claims are false, then one needs an *explanation* of why they are so widely believed; just as, if there are no witches, we want to know what caused people to believe there were. David Hume in Britain and Baron d'Holbach in France in the eighteenth century began the attempt to explain the origin of religion naturalistically. They argued that religion is a response to fear of the unknown, coupled with superstitious attempts to control or propitiate unseen powers.

But why does religion persist in the modern world, now that we understand natural causes? The explanations here come from Westphal's second variety of atheists, the masters of suspicion. Karl Marx, Friedrich Nietzsche, and Sigmund Freud practice the hermeneutics of suspicion, the "attempt to expose the self-deceptions involved in hiding our actual operative motives from ourselves, individually or collectively, in order not to notice . . . how much our beliefs are shaped by values we profess to disown."[21] These three develop their suspicion with primary emphasis, respectively, on political economics, bourgeois morality, and psychosexual development, but each also subjects the religion of Christendom to devastating critique.

Two further steps were needed to make atheism a truly viable position. It would be possible to say that religion may be an illusion, but a harmless or even beneficial illusion in that it shores up morality. So two sorts of arguments were needed. One sort was to show that religion did not serve to reveal anything about the moral order that we could not get just as well by the use of human reason, and I have commented above on the disastrous outcome of this move. Most of the work in philosophical ethics during the modern period had this as its aim. The other was to adduce historical evidence to the effect that religion has, in fact, promoted the worst evils in history.

So within the space of two and a half centuries, roughly from 1650 to 1890, unbelief has become a live possibility. What I intend to emphasize, though, is that this is not merely the excision of God from an otherwise common worldview, but rather the slow development of a *rival* tradition alongside the various theistic traditions and subtraditions. Baron d'Holbach, writing in the eighteenth century, could be called the father of modern atheism. His 350–page *System of Nature* was a systematic avowal of materialistic atheism: a treatment of the world as a whole, humanity's place in it, human immortality, the structure of society, the relation of morality and religion, all as a coherent totality. After circulation of this book, full unbelief became a discussable position for the first time in modern Europe.[22]

I have mentioned some of the crucial developments that made atheism tenable: an alternative account of the origin of religion, first by philosophers and later by anthropologists; theories explaining the persistence of religion in our own day, Freud, Marx and Nietzsche offering some of the most powerful; and the development of accounts of the grounding of morality in something other than the will of God.

Many of Dawkins's publications can be seen to fall within a genre that attempts to contribute to and update this tradition. He offers a new account of the persistence of religious belief that is less suspicious than Freud's, but (I might suggest) less sophisticated as well. Freud's explanation is based on his estimation of human life as constant conflict between the individual's most powerful drives and the worlds of both nature and culture. Culture demands restrictions on impulses, and nature ultimately destroys us through sickness, aging, and death. Religious doctrines are illusions, beliefs induced because they fulfill deep-seated desires. The desire is for an all-powerful and benevolent father who will compensate us in another life for the permanent internal unhappiness that we experience in this one.

According to Dawkins, religious belief held into adulthood is a function of the person not getting over the necessary gullibility of children that allows them to be apt learners, combined with the tendency children have to rigidly retain the lessons that have been drilled into them.

Freud and Dawkins are both influenced by James Frazer's thesis about the origins of religious ceremonies. Frazer traces them to symbolic or representational thinking wherein causal connections are expected between things that resemble one another. While Dawkins merely repeats Frazer's thesis, Freud's very complex theory involves insights into the way believers acknowledge small sins and atone for them as a means of hiding from themselves their deep and total sinfulness.

So what of the epistemological status of religion? I have suggested that Dawkins is a contributor to the modern materialist tradition that is a rival to the various contemporary theistic traditions. The apologetic task is not one of finding proof or evidence for the statement that God exists. It is rather that of finding criteria for judging between rival traditions, each with its formative texts, its account of ultimate reality, and of what that account means for human life.

The task for the Christian apologist then is to show that some version of the Christian tradition stands up favorably against its most significant rivals. One of the major crises facing Christianity in the modern era was the rise of empirical science. Not because of the "Galileo affair" or Darwin, but because of the new epistemology and especially the problem of divine action created by the picture of a mechanistic world governed by deterministic laws. Immanuel Kant and Friedrich Schleiermacher "solved" the problem for liberal Protestants by arguing for a view of theology and religion as too unlike science to possibly conflict, but this was just to put off the hard work of creating a new worldview that integrated the Christian tradition with the new science. The work of theologian-scientists such as Peacocke, harmonizing theology with the scientific worldview, is one of the most important tasks of our day.

Response 8

Arthur Peacocke on Method in Theology and Science and His Model of the Divine/World Interaction: An Appreciative Assessment

Robert John Russell

The interdisciplinary, intercultural, and inter-religious field of theology and science has been growing exponentially since the 1950s, due in large measure to the outstanding contributions of one of its key pioneers: Arthur Peacocke. Building on a prestigious record of achievements in the natural sciences, most prominently physical chemistry, Peacocke's major contributions to theology and science span works written in the 1970s, such as *Science and the Christian Experiment*[1] and the Bampton Lectures, *Creation and the World of Science*[2]; the 1983 Mendenhall Lectures, *Intimations of Reality*[3] and *God and the New Biology*[4]; and the 1993 Gifford Lectures, *Theology for a Scientific Age*.[5] Along the way he has contributed crucially important essays in the CTNS/Vatican Series on scientific perspectives on divine action[6] and the scholarly journal *Zygon: Journal of Religion and Science*.[7] Moreover, in 1983 he wrote a scientific survey of recent developments in physical chemistry and their importance to the evolution of biological complexity.[8]

While these academic accomplishments would surely have secured a lasting place for Peacocke in the history of theology and science in the twentieth century, his leadership in such organizations as the Ian Ramsey Centre at Oxford University, the European Society for the Study of Science and Theology (ESSSAT), the British Forum on Science and Religion, the Institute for Religion in an Age of Science (IRAS), the Zygon Center for Religion and Science (ZCRS), and the Center for Theology and the Natural Sciences (CTNS)—where I am proud to remember Arthur as our third J. K. Russell Fellow (1986)—exemplifies Peacocke's vital role in developing the institutions that were created in this crucial period, institutions that now regularly engage and sustain the dialogue. Most prominent in my opinion, however, is his crucial role in helping to create a new religious order, the Society of Ordained Scientists (SOSc).[9]

This singular and lasting achievement testifies to Peacocke's commitment to the worshipping community as the ultimate context for serious theological reflection on the natural sciences and his particular vocation to the spiritual nurturing of fellow Christians who are practicing scientists.

A full-length book would not suffice for an adequate assessment of Peacocke's work; a short essay such as this can only fall very short of even a reasonable start. Nevertheless I will see how much can be accomplished in cursory form by focusing on just two areas of his work: methodology in theology and science, and divine action in light of science. I have chosen these areas because, following the invitation to this essay, they show both my indebtedness to Peacocke's work *and also* my differences with it. Let me simply state, however, that the surpassing accomplishments of his lifetime of work, all of which have been woven into my writings in one way or another, include Peacocke's aggressive arguments against reductionism and his own particular articulation to critical realism; his robust support for a Christian appropriation of evolutionary biology ("theistic evolution" via continuous creation) with its roots in Anglican theology; his immediate and visionary appreciation of the scientific evidence of "fine-tuning" (or the "Anthropic Principle") and its potential theological significance for creation *ex nihilo*; his early introduction of feminist metaphors for God as Creator; his acute theological assessment of sociobiology even while resisting its overbearing reductionism; his long-standing articulation of panentheism through the slogan, "transcendence-in-immanence," and through his focus on a philosophy of emergence; his deployment of kenotic theology, the theology of the God who suffers with the suffering of humanity, to include the suffering of nature throughout the evolution of life; his use of emergence to warrant theology's claim to the irreducibility of the concept of the resurrection of Jesus of Nazareth against a strictly psychological account of the experience of the disciples; and his candid acknowledgment that Christian eschatology is radically challenged by the far-future predicted by scientific cosmology. For these scholarly accomplishments and for his commitment to the living faith and its gathered worshiping communities, let alone for his mentoring of my career and our deep friendship, I am and will be ever grateful.

Now to the task at hand: two areas where I both follow and diverge from the approaches taken by Peacocke.

Peacocke's Pivotal Contribution to Methodology in Theology and Science: Language as Metaphor and a Hierarchy of Emergent Levels

The methodological framework which has played an essential role in making possible the rapid growth in theology and science is based on four distinct but interrelated claims: first, all language (e.g., both scientific and theological) is metaphorical. Second, the meaning of truth is reference, and its warrant is first of all correspondence; lacking correspondence, truth can be seen as coherence; and finally, lacking these, as utility. Third, the disciplines of the secular academy form an epistemic hierarchy, which can be extended to include theology and ethics. Fourth, the methodologies in the sciences and in theology can be, and should be, analogous. These four claims were initially developed in the pioneering writings of Arthur Peacocke[10] and Ian Barbour.[11] Here I will focus on the first three claims because this is where Peacocke's contributions to methodology seem to me most impressive.

Language as referential metaphor

Many scholars in theology and biblical studies—and many scientists—presuppose a strict dichotomy between scientific language as literal and religious language as metaphorical in the sense of being nonreferential.[12] This dichotomy, if true, would undermine the possibility of an exchange of ideas between theologians and scientists. Beginning with his Mendenhall lectures in 1984, Arthur Peacocke has argued instead that both theological and scientific languages are metaphorical in a rich and nuanced sense that irreducibly includes referentiality and that the meaning of truth in both fields is primarily referential adequacy. Models in both science and theology, and the metaphors out of which they are constructed, are "candidates for reality":

> [They] are reformable and are as close as we can get to speaking accurately of reality: they are not literal pictures, but they are more than useful fictions. They are both representations . . . of aspects of reality that are not directly accessible to us. . . . Christian believers take their models to depict reality, otherwise they would be affectively and personally ineffectual and inoperative. Yet the reality such believers seek to depict is one that the creature cannot claim to describe as it is in itself. . . .[13]

He then draws on a variety of scholars to give a series of arguments in support of this "critical realist" view of theological and scientific language, including Ian G. Barbour, Max Black, Mary Hesse, Sallie McFague, Ernan McMullin, Ian T. Ramsey, and Janet Soskice.[14]

Peacocke deploys a series of metaphors which creatively fire our theological imagination. God is like a radar which sweeps the world to pick out novel events and processes. God "rings the changes," an image of God sounding all possible sequences within the potentialities God built into nature to bring about by continuous creation the truly new. The world is the womb of God

who, like a mother, births her child from within her own being. God is immanent to us precisely as the transcendent God as captured by the metaphor
"transcendence-in-immanence." God suffers with creation by taking the suffering of the world into the divine nature through *kenosis* (self-emptying).[15]

I have used Peacocke's understanding of metaphor through a series of essays
that explore the ways metaphors both shape our exploration of the unknown
in terms of an analogy with the known and as a linguistic bridge between
theology and science. I first met Peacocke at an IRAS/Star Island conference
in 1982. The theme of the conference was the potential relationship between
thermodynamics and the problem of natural evil (e.g., suffering in nature). In
"Entropy and Evil"[16] I constructed a metaphorical relation between entropy in
thermodynamics and both "natural evil" and goodness in nature. Within the
metaphor there lies an analogy between such biological realities as suffering,
disease, death, and extinction and, underlying each of these processes, the
physical processes governed by thermodynamics—in particular, entropy. But
I also pointed to the "disanalogy" (the "is not" of Ricoeur and McFague[17]) in
which entropy contributes to and underlies those processes we understand as
part of natural goodness. Because of this "double role," the ambiguity inherent
in moral good and evil can be seen as prefigured, even in quite muted form,
in the physics that underlies the processes of good and evil.[18]

An epistemic hierarchy

Peacocke has mounted a series of robust arguments against epistemic reductionism dating back at least to his 1976 article in *Zygon* and continuing
through his Gifford lectures up to the present. His approach in its generic
form—that there are increasing levels of quasi-irreducible complexity in
nature that emerge in time through the processes of evolution and that are
studied through a hierarchy of semi-autonomous disciplines—can be found
elsewhere in the literature on theology and science. However, he developed
this approach in a unique way that I have found very fruitful. The *locus classicus* is his Gifford lectures, where he presents in diagrammatic form a "hierarchy of disciplines."[19]

Here Peacocke argues that the sciences and the humanities, including theology, can be located in a series of epistemic levels that reflect the increasing complexity of the phenomena they study. In this epistemic hierarchy, lower levels
along the vertical axis place *constraints* on upper levels (against "two worlds"),
but upper levels cannot be reduced entirely to lower levels (against "epistemic
reductionism"). Thus, physics at the lowest level places constraints on biology
and neurophysiology, found on increasingly higher levels. On the other hand,
the processes, properties, and laws of the upper level (e.g., biology) *cannot be*

reduced entirely to those of the lower level (e.g., physics). This claim accomplishes two moves, which are essential to creating intellectual space for "theology and science": first, it allows Peacocke to move beyond the "two languages" views found not only in Karl Barth but in the diverse schools of Protestant neo-orthodoxy and existentialist theology. These theologies dominated the first half of the twentieth century and precluded any serious interaction between theology and science by keeping them in "watertight compartments." Second, this claim also allows Peacocke to defend the integrity and semi-autonomy of theology against reductionist claims that seek to translate it entirely into psychological language. In my opinion Peacocke is striking out against the vector of thought that leads from Bultmann to the Jesus Seminar by arguing that:

> This concept or these concepts of "resurrection" thus do not need to be reducible to any purely psychological account. The affirmations of the New Testament that propose this concept can properly be claimed to be referring to a new kind of reality hitherto unknown because not hitherto experienced. "Resurrection" manifests a new kind of ontology in the nature of the risen Jesus. Such a proposal illustrates the general thesis of this essay: that in theological discourse about experiences of God and of divine action there is a parallel to those processes whereby emergent realities are apprehended in the hierarchy of complex systems studied by the sciences and so are given at least tentative ontological status.[20]

My own approach to the resurrection of Jesus is to argue that something even more *sui generis* is involved than that which is possible within the framework of the emergence of irreducible processes in nature.[21] Nevertheless I applaud Peacocke for his development of a philosophy of emergence in general and his argument against the reductionisms and isolations of our competitors.

The horizontal axis of his diagram is also important: it ranks natural phenomena in terms of their increasing size within the same epistemic level. Thus physical systems at the bottom of the diagram range from elementary particles to galaxies as studied by fields ranging from physics to cosmology. I find this one of the most important and yet underappreciated aspects of his epistemic scheme,[22] for it makes abundantly clear that cosmology is a part of physics and as such places constraints on all the supervening disciplines, including theology. Thus, unlike some scholars who fail to understand the challenge to Christian eschatology posed by the far future in Big Bang cosmology, I believe that those scholars who do view the bodily resurrection of Jesus as a proleptic event of the transformation of the universe into the New Creation must face this challenge squarely and exhaustively.[23]

In my view another crucial methodological step toward making theology and science possible was offered by Ian Barbour. Here, too, we see a "levels" approach to complexity in nature, but Barbour adds to this the insight that within this hierarchy, each level involves similar methods of theory construction and testing in both science and theology.[24] In my own work I have combined these two ideas from Peacocke and Barbour—Peacocke's arguments for an epistemic hierarchy and Barbour's stipulation of an analogy between the methods of science and of theology—into an overarching proposal represented schematically by Figure 1 (below). Here, following Peacocke's argument for a hierarchy of disciplines with their inbuilt rules of constraint and emergence,

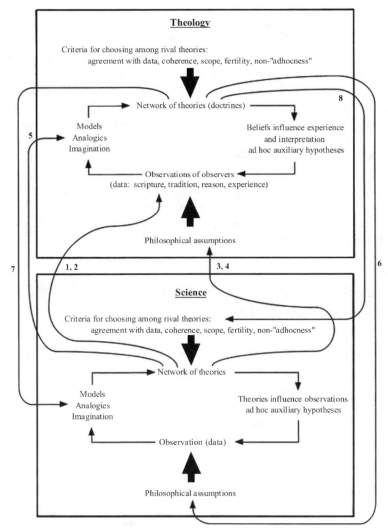

Fig. 1: The Method of Creative Mutual Interaction (CMI)

I have placed Barbour's figure for theological methodology above his figure for scientific methodology. (Here for simplicity I have limited the conversation to physics and theology.) This figure incorporates Barbour's analogous methodologies, but now ordered according to Peacocke's epistemic hierarchy.

With this diagram we can now clearly distinguish between the diversity of ways scholars actually seek to relate theology and science, clarifying the differences in their approaches, and thereby clearing away a great deal of controversy over what should be the "only way" to relate them, while assessing the value of each on its own terms. We can also explore new ways to relate them that have received much less attention than the standard ones. I do so by tracing eight paths that represent these different ways in the diagram: more routine paths lead from science to theology while others, more controversially, lead from theology to science.[25] Together these eight paths portray science and theology in a truly dynamic, though asymmetric, interaction. Moreover, the entire set of paths taken together represents an overarching view of the theology/science discussions that no one path can adequately suggest. I call this methodological proposal "the method of creative mutual interaction." It is first of all a description of the ways scholars allow science to influence theology and vice versa. But it is also prescriptive: I believe a more intentional exploration of such influences could be fruitful for science as well as theology. Most of all, it could be particularly fruitful for "theology and science" because it allows us to delineate the conditions and criteria for real progress in research on these topics.[26] This methodology would not have been possible without the proposals of Peacocke and Barbour, to whom I owe tremendous gratitude.

Peacocke on God's Action in the World via the "World-as-a-Whole" Model

Peacocke has developed a variety of models of God's interaction with the world within the framework of panentheism, with its stress on divine immanence as well as divine transcendence.[27] These models combine "top-down" and "whole-part" approaches in fascinating ways. Peacocke begins by challenging the mechanistic view of nature which, because of its underlying acceptance of Laplacian determinism, led to an interventionist understanding of God's actions.[28] Here, for God to act in specific events in nature, God would have to violate the causal processes of nature and God's action, in turn, would contradict the laws of nature that science uses to describe natural processes. Instead, Peacocke urges us to see God as being in "continuous *interaction* with and a continual influence on the created order . . . (rather than) 'intervening' in any way disruptive of the natural and human processes he has created."[29] He then draws on contemporary science for evidence to support his challenge to mechanism.[30] Here fields like quantum mechanics and non-linear dynamics seem to point to "permanent gaps in our ability to predict events in the natural world." This might, in turn, lead to a new view of nature as ontologically open and thus open the way for a new approach to divine action in the world. For this to be achieved, Peacocke raises two key questions: (1) Even if we cannot predict the future, can God know it? (2) Does the new flexibility of nature provide room for God to act without intervening in the causal regularities of nature?

First, Peacocke addresses these questions from the perspectives of promising new foci in classical physics, namely the many-body problem, chaos theory, and non-linear, non-equilibrium thermodynamics. Some scholars have found these new fields to offer a promising approach to non-interventionist divine action. However, since they are still in principle a part of classical physics, Peacocke's prognosis is negative. On the one hand, it is true that the future state of these systems is, for us, unpredictable in principle, not just in practice. Chaotic systems, for example, exponentially amplify tiny, undetectable fluctuations in the initial conditions into different future states of the system that are for us unpredictable.[31] But does this unpredictability hold for God as well? No, replies Peacocke. Unlike us, God can know the future state of systems like these by predicting the future from the present (Question 1) because these systems, being essentially classical in nature, are determinist, and if God is omniscient, God's exact knowledge of the present would allow God to predict the future.[32] Moreover, for God to act to alter the future (Question 2), God would have to change the initial conditions of these systems. Although this kind of divine action would never be detectable by us, it would still be a form of intervention. Thus Peacocke—in my view correctly—rejects chaos theory, etc., as offering a non-interventionist approach to divine action.[33]

Next Peacocke turns to quantum mechanics in his search for a non-interventionist account of divine action. He acknowledges the possibility of a deterministic interpretation of quantum uncertainties (i.e. a "hidden variables" view) but favors the argument that quantum mechanics points to ontological indeterminacy in nature, which he reports as the "majority view of physicists."[34] He cites radioactivity as an example of such genuine indeterminacy in nature.[35] In response to Question (1), Peacocke views quantum indeterminism as a limitation even on God's omniscience; because of quantum indeterminism, God can only know the future probabilistically. Moreover, since God has created the world this way, this limitation is "self-limited." According to Peacocke, this entails an answer to Question (2) as well. Since "God cannot know precisely the future outcome of quantum dependent situations, [God] cannot act *directly* to influence them in order to implement the divine purpose and will."[36]

Unlike my endorsement of Peacocke's critique of the appeal to new foci in classical physics as discussed above, my response to these claims about quantum mechanics is mixed. (a) In my opinion the indeterministic interpretation of quantum mechanics does not imply that God cannot know the future. Instead, quantum indeterminism only implies that God cannot foreknow the future based on God's knowledge of the present just prior to a quantum event—say the decay of a radioactive atom. But within classical theism God knows the future not by foreknowledge based on God's knowledge of the present but by knowing the future in its own present actuality. This claim is not challenged by quantum indeterminism. (b) If God acts within nature (though never as a natural cause) to provide the sufficient cause of a particular quantum event, then God can foreknow the future since it is determined by the way God chooses freely to act with nature to bring about that quantum event.

Third, Peacocke targets complex processes in nature which display both "top-down" and "bottom-up" causality. On the one hand, events at "lower" levels of complexity influence the behavior of the total system in a "bottom-up" way; yet on the other hand Peacocke claims that the system as a whole supervenes on, and thus constrains, the behavior of its lower levels, though in a way consistent with the laws describing these levels. Now he makes his key move: by applying this notion to the universe—which he calls the "world-as-a-whole"—we can think about how God interacts with the world to bring about particular events in nature while respecting and not violating the law-like processes of the world.[37] The concept of panentheism again helps Peacocke bring together God's transcendence of the "world-as-a-whole" with God's immanence "in, with, under and through" natural processes as suggested by his fascinating term, "transcendence-in-immanence"[38]:

> Attempts to hold together the notions of God's transcendence over the world, his ultimate otherness, with that of his immanent Presence in, with and under the world often find it helpful to deploy models of the world as being, in some sense, "in God," but of God as being "more than" the world, and so as the circumambient Reality in which the world persists and exists ("pan-en-theism").[39]

The key, however, is Peacocke's focus on God's causality in a "top-down" manner[40], which does not involve God intervening in the laws and regularities of the lower levels. The crucial piece that makes Peacocke's approach work is his conceptual model of the "world-as-a-whole;" through it, Peacocke can speak of God not only sustaining the world but interacting continually with it to achieve both general and particular effects.

I am in solid theological agreement with Peacocke here in his understanding of God's relation to the world, and I admire the way it offers him a unique approach to our understanding of God's interaction with the world. The former is grounded in the rich theological tradition that emphasizes the radical ontological distinction between God and "all-that-is," while at the same time it insists that God is intimately present to "all-that-is." The latter is a splendid attempt to achieve the goal of a non-interventionist account of divine action in a unique way: not by `identifying domains of ontological openness within nature as I do through a quantum mechanics–based non-interventionist approach to objective divine action ("QM-NIODA"), but by viewing God's action as directed to the world-as-a-whole. Divine action conceived in this way achieves particular effects as indirect and mediated by God's direct action on the whole rather than "within" it.

The question I want to raise, however, is whether the concept of "the world-as-a-whole," which Peacocke uses to explore both the theological understanding of God's ontological relation to the world and a "top-down" non-interventionist account of divine action, makes sense when we try to interpret this concept in terms of contemporary science.[41] My worry is that it does not, and this worry might best be expressed by turning to the diagram which Peacocke used to illustrate God's interaction with the world in the Gifford lectures and in his chapter in the CTNS/VO volume on *Neuroscience and the Person*, although its legend is not included as such in the present essay on "A Naturalist Christian Faith."[42] (See Appendix C.) The figure is meant to suggest both God's transcendence to and immanence within the universe as "the world-as-a-whole."

In the figure a dashed circle represents the boundary of the "world-as-a-whole." Within the circle we find all of creation, including a symbol standing for humankind at the center of the circle. God as the transcendent Creator is then symbolized as coming in from the infinite region outside the circle and acting on it. The effects of that action are transmitted by natural causes into the interior of the circle, eventually reaching humanity. God is also symbolized as acting immanently on all points of the "world-as-a-whole." Here God's action is orthogonal to the paper on which the diagram is inscribed, as indicated by the "q" at the center of the diagram.[43]

Now Peacocke is quite clear that a figure such as this can only go so far in depicting the God-world relationship since God's relation to creation is not "spatial." Indeed, spatiality is one of the properties of the created world. As Peacocke rightly insists, the figure is instead meant to represent God's ontological relation to the world as well as God's interaction with the world:

> Figure 1 attempts to express on a two-dimensional surface the relation between different modes of being—it is an ontological representation (rather like a Venn diagram representing the logic of the relation between different classes).[44]

So far we are on solid ground. My concern, however, is with the extent to which the figure relates to what science tells us about the universe, that is, "the world-as-a-whole" in the framework of science. On the one hand, if the dotted circle merely represents the ontological distinction between God and the universe, then it leaves little room for the implications of science. After all, theologians have been making this same strictly ontological distinction between God and creation in eras dominated by the vastly different cosmologies of Aristotle/Ptolemy, Copernicus/Newton, and now Albert Einstein. If, on the other hand, the dotted circle is also meant to represent what science tells us about the universe, if in particular the dotted circle is meant to represent the limits or the boundary of the universe, it fails to be a valid interpretation of what science tells us about the universe—in a quite remarkable way. But if this is so, and I believe it is, then it also fails to be a valid interpretation of God's relation to the universe *in light of science*.

The reason for the failure is based on the "counterintuitive" fact that, according to contemporary scientific cosmology, *the universe does not have a boundary*—it is unbounded.[45] It is relatively easy to understand this using the basic models in Big Bang cosmology.[46] In these models the topology of the universe is a three-dimensional hypersurface expanding in time. The hypersurface is either "spherical" (closed/finite in size, eventually recontracts), "flat" (open/infinite in size, expands forever), or "hyperbolic" (open/infinite in size, expands forever). The spherical three-dimensional hypersurface, for example, is analogous to a two-sphere such as a beach ball (i.e., the two-dimensional surface of a beach ball, not its three-dimensional volume). Obviously the surface of a beach ball has no boundary; an ant crawling on it never comes to an "edge," although in time it might return to its starting point.

The universe, then, is simply not analogous to Peacocke's disk circumscribed by a dashed line. There is no boundary to the universe. Therefore there is no boundary on which God can act from "outside," and thus no top-down or whole-part divine action on the universe as a whole when the universe is understood in light of science. As long as the theological claim about the relation between God and the world is strictly ontological, there is no "boundary" problem, but then science is *irrelevant* to this theological claim. As soon as science comes into the picture at all, the dotted circle, and thus the figure as a whole, fails to represent what science tells us about the universe.

But wait, there is still one way that might take us out of this cul-de-sac. One might say that "the world-as-a-whole" is best thought of as the entire three-dimensional curved hypersurface of the universe considered at a specific moment of proper time J (i.e., the age of the universe) and not its boundary (of which there is none). God's action might then be thought of as immanent everywhere and at every point in the hypersurface (i.e., as indicated by the "q" symbol). Ironically, given Peacocke's dislike of a quantum mechanical approach to divine action, this actually moves the conversation about divine action to precisely this approach: divine action at every point-like quantum event throughout the universe. Perhaps God's "top-down" action on every part of the universe is intimately related to God's "bottom-up" action "in" every part of the universe. Just as Peacocke gave us the phrase "transcendence-in-immanence," we might name this idea "top-down-through-bottom-up."

Conclusions

Arthur Peacocke is truly one of the key pioneers in the field of theology and science. His contributions span multiple intellectual discussions and move beyond them into the broader context of the living community of scholars internationally who have benefitted from his vision and scholarship. He will hold a place forever in the history of this new and exponentially growing field thanks to his integrity, his candor, his openness, his personal faith, and his warm friendship. For his mentoring of my career and our deep friendship and shared faith in Christ, I am and will be ever grateful.

Personhood, Spirit, and the Supernatural

Keith Ward

Arthur Peacocke has what I think is the unique distinction of having senior Doctorates of the University of Oxford in both Science and Divinity. These doctorates are only awarded for work of a quality "very substantially higher than that required for a Ph.D." So he needs no commendation from me. Nevertheless, it is only right for me to say that he is a seminal figure in the field of science and religion, and that his theological contribution ought to have completely restructured the study of theology at least in the United Kingdom. It has partly done so, and to the extent that it has not, the fault lies with those theologians who are afraid of making fools of ourselves by trying to talk about science. In general theologians are very good at biblical studies, at the history of doctrine, and at ancient languages. But we tend to be rather bad at learning to make straightforward, informed, intelligible statements about modern scientific knowledge and its implications for theological understanding. Peacocke's work should make it easier for us, and I hope that modern theology will be able to use his work to help restate its major claims in a more coherent way.

In this short chapter I want to discuss the question of miracles. Miracles have been central to Christian faith in the past, though theologians like Friedrich Schleiermacher, in the late eighteenth century, felt able to dispense with them. I myself approve of Schleiermacher's dictum that Christians should only accept the occurrence of those iblical miracles that they feel to be well-enough attested to be accepted on the testimony of the biblical writers. But the question I want to ask is the more general one of whether an account like Peacocke's rules out a fairly strong sense of miracles or not. He sometimes implies that it does, but overall the situation is not quite so clear, and I wish to probe it a little.

Are There Supernatural Forces?

Peacocke calls his a "naturalistic theology," and speaks of "theistic naturalism." This is at once a term that will make any personal theist, who believes that God is thoroughly supernatural, wince. But one needs to look more closely at what he means. The Oxford English Dictionary defines "naturalism," in its main relevant sense, as "the belief that only natural (as opposed to supernatural or spiritual) laws and forces operate in the world." So it seems to be a causal theory that excludes any causal influence of a supernatural being upon events in the physical world. It does not necessarily, however, exclude the existence of a supernatural being. But can a theistic view really rest content with saying that there are no supernatural forces operative in the world at all?

Peacocke does believe that God, a supernatural being, has caused, and continues to cause, the whole universe to exist. He even says that the universe is "God's act," thus implying that the universe has an intentional cause, that it is brought about for a purpose by a conscious being. God is a consciousness that envisages the world and actualizes it for a reason. How plausible is it, then, to say that such a God will refuse to operate in the world in particular ways? For that is what "naturalism" seems to imply. There could indeed be such a God, but Christianity, at least, speaks of particular divine acts—especially the incarnation, the resurrection, and the descent of the Spirit on the early disciples.

There may have been no virginal conception. But if one speaks of Jesus as in any way uniquely sinless or having unique knowledge of God, the most intelligible way to think of this is to say that God has caused Jesus to be sinless or knowledgeable, that God has operated within history to raise this human person to a unique relationship with God. If the resurrection is truly the manifestation of a living Jesus to the apostles after the death of his body, this again is most intelligibly seen as a particular divine act. And if one believes that the Spirit is active in human lives, once again this seems to commit one to some divine particular causality. My question is, does Peacocke's account allow for this or not?

Peacocke is an overt proponent of objective theism, by which I mean that "God" is not a word that expresses or refers to some personal attitudes or states of mind. It refers to a reality that would continue to exist whether or not there are any persons, or whether or not there is any universe. Further, God is a spiritual reality, not having material location or extent but possessing consciousness and intention, knowledge, and will. These terms may be analogical; they may refer to something in God very different from anything in human minds. But the terms are felt to be appropriate, and they entail that the whole universe exists as a result of a conscious decision on God's part. This is a massive dualism. God is distinct from and prior to anything material. I believe that

Peacocke has no problem with that sort of dualism, between a spiritual God and the created universe. But he does want to deny any dualism of spirit and matter *within* the cosmos, or any form of particular spiritual causality within the cosmos. I just wonder whether he needs to make such a strong denial.

What Is Panentheism?

In his elegant and illuminating Essay he espouses "panentheism," the view that, in some sense, the world is in God or is the body of God. But to me the natural implication of such a view is that the body does not run in accordance with a set of completely autonomous and internal laws. The body does, to a large extent, what the mind tells it to do. The mind, many of us would think, actually has a causal influence on the body, and in pretty obvious ways. I would therefore expect a panentheist to see supernatural causality (the influence of the spiritual reality of God) virtually everywhere in the physical cosmos. At least some physical events should be identifiable as purposive acts of God. If they are not, the case for panentheism seems to be greatly weakened.

Panentheists hold that God and the universe together form one being, rather as mind and body form one being in humans. For most Western philosophers, this analogy falls at the first fence. For human minds depend upon human bodies and brains, and they are largely, if not completely, caused to be what they are by events in those bodies and brains. But the divine mind does not depend upon the universe for its existence. God, on some views, may depend for some of the particular contents of the divine consciousness on what happens in the physical universe. But the divine consciousness would exist without any universe; so this seems like a causal relationship, rather than a part-whole relationship. God may act in the physical universe, but God is not limited by the aging or decay of the universe. To put it sharply, the physical universe contains nothing like a brain that interacts with the divine mind.

What, then, does panentheism come down to? It seems to come down to saying that no part of the physical universe is ever absent from God—in other words, the perfectly traditional view that God is omnipresent. God and the universe are not two substances of the same sort. Yet that is exactly what dualism says: spirit and matter are not the same sorts of thing.

Perhaps, then, panentheism is making the point that at least God is affected in some way by the universe. But an equally good way of making this point is to say that God is a truly personal reality who causally interacts with the universe, being affected in knowledge by what occurs in the universe, and responding by acting in particular ways within the universe. The personal analogy can cope with the idea that human persons are truly other than God, with their own wills and experiences. They are not just parts of a super-organism. In short, God is a purely spiritual reality who interacts with the universe in quite a strong sense, who knows it perfectly and can act upon it directly. But God retains an essential otherness from both material reality and finite persons, with whom God can have a relation as between two distinct realities (a purely spiritual reality and many embodied spirits). God is the only self-existent personal reality who interacts with the universe and with dependent persons within it. Indeed, God creates the universe by an intentional act. Since the bringing of something into being is the strongest possible causal influence upon it, it may seem unnecessary to espouse a theory that denies any particular causal influence of God upon the universe.

Peacocke is, of course, well aware of this, and he wants a more restricted sense of the term "naturalism," which would mean only that no being outside the web of cause-effect relations ever *violates* the laws obtaining within that web. There is no dualism, he says, and there are no miracles, in the sense of events that "break" the laws of nature. So he terms his own preferred view an "emergent monism," and insists that God is a non-intervening cause of the universe.

So far, this sounds like the Great Clockmaker of classical deism, who causes the universe but allows its laws to run on their own without any divine supervision or intervention. But I do not think Peacocke really means this.

Law, Causality, and God

It was David Hume who defined a miracle as a violation of a law of nature—and when Hume chose this expression, we may be sure he intended to cast ridicule upon the very idea of a miracle. For what rational Creator would institute a law, only to violate it himself?

The implication of such a view of God is that God intends to have no personal relation to the universe or to anything within it. A personal relation is one for which a person knows that another person exists and apprehends them as existing. Further, one person acts upon the other in such a way as to change their knowledge or feeling. This action is usually mutual, such that persons exchange information or cooperate with one another so that there exists a certain sort of causal reciprocity between them, one causing certain mental states in the other and the other responding by causing mental states in the first. It is in such causal reciprocity that personal relationships normally consist.

That seems to be ruled out in Peacocke's account, since God never exercises a *particular* causality upon anything in the universe, including any persons it may contain. God never directly causes anyone to have knowledge of God or to have new mental states as a result of such knowledge. Nor can anyone cause any change in God.

People may come to knowledge of God, in accordance with the general laws God has placed in the universe. But any such knowledge must be simply a result of ordinary causal laws, not of any actual approach that God makes to persons.

This is not totally unusual for theology. The God of classical theism, it may be said, is immutable, and so does not have fully personal relations with finite beings. An immutable God must put into the universe all the actual events that ever happen within it, in one and the same timeless act. God cannot truly respond to things that creatures do, except in the sense that God timelessly decrees both what they will do and how God will respond on that occasion. The decrees of God are timeless, even though it may seem to us that God hears what we say (in prayer, for instance), and only then decides to respond. The truth is that God eternally decrees both what we say and the divine response, in the same timeless single act that is the act of creation.

If this is the classical view, it does not actually seem too different from Peacocke's. God is changeless, and does not act in time, in response to the acts of creatures—though it may seem that way to them. So God is not interfering with any laws of nature.

But there is still a problem. Suppose God sets up the whole universe, from beginning to end, in one timeless act. Will God decree that everything that happens in the universe will happen in accordance with laws that allow no supernatural interference or addition?

This immediately raises the question of what is meant by a "law of nature." On one widely held view, there is a finite set of laws that governs the relations of fundamental particles (superstrings, perhaps), and those laws operate universally (everywhere) and without exception. There are other laws for biology and psychology, for instance, but in the end they all reduce (perhaps not conceptually, but ontologically) to the basic level of physical laws. We can now suppose that this set of laws wholly explains everything that will ever happen in the universe, even though it may be too complicated in practice for us ever to test that supposition (how could we know that nothing will ever happen that the laws cannot explain?).

That is a coherent supposition. But Peacocke explicitly denies that is his view. He is an "emergent monist." There are layers of reality, such that the wholes cannot be explained solely in terms of the parts. He points out the difference between epistemological non-reducibility and ontological non-reducibility. The former makes psychological explanations, for instance, simply a sort of short-hand for whole sets of physical events too complicated for us to note in detail. The latter supposes that there is more than short-hand here. Sufficiently complex and integrated arrays of particles create a situation in which a new sort of reality comes into being. That new reality may depend on the continuation of the integrated array in question, but it is a new sort of thing that exists. Some properties only exist when there are large integrated numbers of simpler physical parts in a specific relationship. It will not be possible to predict the properties of the whole just from knowledge of the properties of the parts. Unless, that is, it is one of the causal laws that complex arrays of simple particles will produce a new property. That could indeed be a causal law implanted from the first moment of creation by God. But if so, it is a "hidden" property that is never exhibited until the complex array comes into existence, and that is not even knowable until it does so. This throws some doubt on the idea of "laws of nature," for they will be indefinitely open-ended. We will never know what they are until we see all the possible emergent states of things in suitably complex situations. The idea of a finite and exhaustive set of laws of nature collapses with the thought that some laws can never be known until the history of the universe is completed. In that case, the complete set of laws of nature will just be a redescription of all that has happened in the history of the universe. But it will carry one bold assumption, which is that everything that ever happens can be brought under some law, however complicated, a law that says, "Whenever x happens, then y will happen." But if we will never know all the laws of nature until the end of the universe, how could we ever be justified in making that assumption?

What is the alternative? It is, quite simply, that some events that happen cannot be brought under universal and necessary laws. This alternative is not some peripheral and obscure fancy. We are totally familiar with it every day. Social scientists have sought for laws of social interaction in vain. If humans have responsible choice, their actions will not be all subsumable under laws. If I write a book, I may never be able to say, "When all these events occur together, this book will (inevitably) be written." I may say, "This book could be written. Its writing can be explained by pointing out all the experiences I have had, events that caused me to write it, and so on. It is not without any explanation. But none of these explanations entails that precisely this book, and no other, in all its specific details, would be written." Personal explanation, in other words, allows for alternatives. If it exists, and is irreducible to physical explanation (what you might call "epistemic non-reducibility"), this strongly suggests that there are a great many events in the universe that are not completely explained just by reference to physical, determining, laws. That does not deny that there are such laws. It just says that they do not cover everything. There are events, and millions of them, that do not fall under universal physical laws.

Do they then "violate" such laws? Do human persons "intervene" in physical nature? That seems absurd. In which case it looks very much as though what we call the laws of nature are deterministic causal relations between identifiable and denumerable items that obtain in situations in which no other causal factors enter into the situation—namely, in laboratories, under controlled conditions, or in abstract scientific models. There are many such situations. But there are also many situations that are not thus, where causal factors exist that may need different laws we do not yet know, or where deterministic laws may be the wrong model to use.

Suppose we ask the question, "How many sorts of causal influence are there at work in the universe?" I think an honest scientific answer would be, "We do not know." Perhaps we cannot know, since we have limited access to the causal roots of being (we know hardly anything, for instance, about dark matter or its causal influences on our universe).

However, if that is so it is due to our limited knowledge. God would presumably know what properties would result from complicated arrangements of simpler parts. The crucial case is, of course, consciousness. Given only properties such as orientation spin, location, momentum and electric charge, how could anyone predict that a sufficiently complex arrangement of millions of items with those properties would produce awareness? It may be a hidden law of nature, known to God alone, that they will do so. It would indeed be very suitable for a law of nature to state that whenever a brain exists, consciousness will exist. That is just the sort of deterministic law we would like, for it makes consciousness non-arbitrary and underlines its dependence on brains. But it is a deeply hidden law, pointing to potentialities in the physical world that none could guess until they happen.

But if such emergent properties are real, perhaps, Peacocke suggests, "to be real is to have causal power." We would need causal laws that describe how consciousness affects lower levels of matter. And perhaps deterministic laws are not the best model at this level. We may need a non-deterministic or probabilistic model of causal influences—just as, arguably, we need such a model for quantum phenomena. So now we have an indeterminately open set of causal laws, some deeply hidden, and some of which are probabilistic. The question is, does this picture enable us to draw a clear line between the natural and the supernatural?

The Cosmic Mind

Such a line could be drawn when we could in theory list all the laws and properties of "nature." But if those properties and laws are indeterminately open, we cannot in principle know what are the limits of whatever sorts of causality there may be. If human minds can have non-determining causal influence, may there not be a cosmic mind that exercises similar sorts of influence?

For a theist, there is such a cosmic mind, but it is not a product of very complex arrays of physical particles and of laws stating that a cosmic mind will come into being in such-and-such circumstances. The cosmic mind is the causal origin of physical nature and its laws. It is not, *ex hypothesi*, caused by anything, but is the cause of everything else.

But now the picture of a Mind bringing about a deterministic and completely lawlike universe begins to lose its appeal. Would it not be likely to bring about a universe in which beings something like itself may come to consciousness and creative activity? In other words, would the laws of nature not be intrinsically goal-oriented, precisely ordered to bring about finite minds? In fact, if the laws of nature are "other things being equal" (*ceteris paribus*) laws, which obtain deterministically only when other causal influences are not present, does that not entail that the laws of nature are always in principle open to the ever-present causal influence of God? If God is an ever-present causal factor, the deterministic picture loses its appeal, except as a default position for physical situations in which God chooses to exercise no specific causal role.

There is no violation or interference, for the universe is in its essential nature an expression of mind, and its laws are abstract schemas of what nature is like when it is not very personal but is serving as the physical condition of the possibility of emergent, developing, self-shaping societies of finite minds. Miracles are not violations of laws of nature, for laws of nature are self-limitations of the personal influence of the Supreme Mind, which are necessary to allow finite, morally responsible, and relatively autonomous minds to exist and grow. A miracle is a withdrawal of this limitation, disclosing in a unique and unrepeated way the personal ground of all existence and something of the direction and goal of its ever-present sustaining influence.

Is this theistic naturalism? Is it emergent monism? Is it panentheism? Is it what Peacocke believes? I think it is at least perfectly consistent with what he believes. For he speaks of Christ as an "an initiative from God," of Jesus as being "wholly open to God," and of particular events beings "taken into the life of God." These statements imply that there are positive personal relationships between, at least, Jesus and God, and by implication between all finite persons and God. God takes initiatives, and receives finite persons into the divine life. That implies both otherness and relational unity. It implies that people can be changed (affected) by their knowledge of God, which in turn is the result of a divine initiative. It would be hard to imagine a more interactionist view of divine-human relations than this.

The Question of Miracles

Peacocke then speaks of "God's non-intervening but specific" influence on the world. God influences reality at all levels, he supposes, but increasingly at higher levels of organisation. So there are particular divine influences on the world after all. If that is the case, I would see Peacocke's case not as being that there are no miracles, no particular divine acts of an extraordinary nature that do not violate laws of nature though they do not fall under laws of nature either. Rather, he would be saying that miracles cannot just be arbitrary interferences in an otherwise elegant and lawlike cosmos. They must have their own form of intelligibility and rationality. In this I think he is wholly correct. God does not break laws of nature, arbitrarily and at random. God and the universe are not two quite distinct substances that have only some sort of external relation to one another. And mind and matter are not two distinct substances contingently connected through the pineal gland or the brain, which could (in the human case at least) exist equally well, or even better, if they were not connected.

What he does show, however, is that there is another way to remedy these misconceptions. There is in fact no closed causal deterministic web of causality, ruling universally and without exception over every physical thing. Laws of nature are abstractions from the very complex causal structure of the physical world, in which we do not know every form of causality there is. But we have good reason to think that there are non-deterministic causal influences at work (what he calls "whole-part influences"), and we cannot set any *a priori* limit to what they are or how precisely they operate.

I see no reason why there should not be miracles, extraordinary disclosures of God, but they will not be arbitrary whims of a completely unpredictable super-person. They will be parts of a rational unfolding of the true nature and goal of the cosmos, in relation to its spiritual origin and basis. They will have intelligibility and order. But they will not come within the purview of the natural sciences, which need to measure, observe dispassionately, and repeat. This, however, is in my view only paradoxically called "naturalism," for it is committed to the priority of the personal and of that which is more than spatio-temporal.

There is another way of conceiving of the relation of God and the universe than as a purely external relation (or, in classical Christian theism, a relation that does not change God at all). This is simply to say that the relation between God and the universe is genuinely mutual, and that the divine Spirit interpenetrates the universe at every point (not via some analogy to a human brain). God is changed by what happens in the world, and God responds, though in ways that respect the order of laws that permit creaturely self-shaping. The best analogy for God and the universe is not that of one mind to one body, but that of a society in which genuine "otherness," genuine relationship, and genuine communion (in which each is changed by the other, but there is unity of purpose and experience) can all exist.

Similarly, there is another way of conceiving the mind-body relationship than as a contingent, almost accidental, and unnecessary connection. Even the much-maligned Descartes said, "I am not just lodged in my body like a pilot in his ship, but . . . am so confused and intermingled with it that I and my body compose . . . a single whole" (*Meditation* 6). Human minds are dependent on the existence of brains, and the functioning of brains depends on the functioning of bodies and of the social environment beyond the body. In humans, mind and body form a single whole; yet it makes sense to speak of brain-states causing conscious thoughts and feelings, and of thoughts causing physical events in the brain. This relationship is peculiarly intimate, and it is proper for mind and brain to be "intermingled" in one whole, so that they cannot be parted without the loss of some essential element. For Descartes, they were logically separable. But a whole human person needs both elements, and it is for that reason that resurrection is preferable to the immortality of a separated soul, bereft of its external means of expression, communication, and of obtaining new knowledge and experience.

A Personal God in an Open Universe

In conclusion, I would repeat that I am enormously impressed by the way in which Peacocke sets Christian beliefs within a thoroughly scientific context and worldview. I am wholly convinced that he has given a remarkably coherent and illuminating exposition of Christian faith within an evolutionary context. I agree with him that God must be reconceived as creating and responding to an open future, that God works through an interplay of chance and law, that God suffers in all creation, and that God influences but does not determine the way things go, inviting us to be co-creators in shaping the future in a truly creative way.

I do think, however, that he can sometimes sound more radical than he really is. When he says he is a naturalist, a panentheist, and a monist, it sounds as if he is ejecting a causally efficacious God from the physical world altogether. Yet he holds that one can have a fully personal God whose causal influences on the world are real and form a fully intelligible pattern without such influences being discernible by the methods of the natural sciences. For such a God, miracles are perfectly possible, though they will have to be rationally integrated parts of the divine purpose for the cosmos. Such a personal God will relate to humans as beings with a proper degree of autonomy and self-shaping, not just as parts of the divine being. God is quite distinct in being from the material universe and all finite persons within it. But God truly relates to events in the cosmos in ways that do not violate its laws, but that make a real causal difference to how things go.

God does not arbitrarily break laws of nature, though laws of nature also do not form a closed and complete deterministic system. God is not just externally related to the world, but is in fully personal and dynamic interaction with an evolutionary universe. Reality is many-layered, so that there are many levels of ontology, not reducible to one another, and conscious knowledge of environment is one of the higher levels. For me, the implications of this profound view of the God-world relation are that it is possible to take a realistic view of at least some miracles—that is, to say that they actually occurred. The body of Jesus could de-materialize in the tomb. The mind of Jesus could be infused with unique knowledge of God the Father by a special divine act. All one would have to show is that such truly extraordinary acts fit into an intelligible pattern through which the laws of the physical cosmos are seen to point towards a supernatural transfiguration of the cosmos into the life of God. No doubt individuals will continue to disagree about which of the biblical miracles fall into such an intelligible pattern and which are mere legendary inventions. But I think that Peacocke's account is compatible with a fairly robust acceptance of miracles, as long as one has a flexible enough view of laws of nature, so that we can avoid speaking of miracles as "violations" of law but can continue to speak of them as particular divine acts within the warp and woof of history, which are not covered by any general laws of nature. If I am right, Peacocke is not quite as frightening to some more conservative Christians as some of them think. Indeed, his view can be heartily endorsed by them as glorifying the creator of such a beautiful, elegant and intricately ordered universe as the one in which we exist. I hope it will be.

Response 10

On Divine and Human Agency:
Reflections of a Co-Laborer

Philip Clayton

Religion is the vision of something which stands beyond, behind, and within, the passing flux of immediate things; something which is real, and yet waiting to be realized; something which is a remote possibility, and yet the greatest of present facts; something that gives meaning to all that passes, and yet eludes apprehension; something whose possession is the final good, and yet is beyond all reach; something which is the ultimate ideal, and the hopeless quest.

—Alfred North Whitehead[1]

Religion will not regain its old power until it can face change in the same spirit as does science. Its principles may be eternal, but the expression of those principles requires continual development.

—Alfred North Whitehead[2]

A set of brilliantly interconnected insights underlies Arthur Peacocke's work. The core motivation for his naturalism is the recognition that science stands in tension with many traditional views of divine action. If one accepts a picture of the world consistent with scientific practice and results, then one cannot imagine that God regularly intervenes in the natural order in a miraculous way, setting aside the patterns of nature and directly bringing about particular physical or chemical changes independent of the causal antecedents for these particular events. Peacocke insists, however, that preserving the integrity of the natural order is not inconsistent with Christian faith. The world can still be created by God and sustained in its existence at every moment by the divine will. Moreover, the world can be located within God, as panentheists maintain, such that God is "in, with, and under" all things, present to them in the most intimate way possible. The result is a notion of divine influence strong enough to undergird many, if not most, of the traditional Christian doctrines. This, at any rate, is my understanding of the program of Arthur Peacocke.

It is always difficult to write a response to the work of one's teacher. In younger years one tends merely to restate the teacher's own views. Then comes a phase when, in order to prove one's own independence, one criticizes overly harshly, hence often unfairly. With age, however, comes a new freedom to recognize the ongoing influence, to acknowledge one's debts with gratitude, and to express misgivings without onus.

The development of my own work in the field of science-religion discourse is intricately linked to the research program of Arthur Peacocke. Reading his work and engaging in intensive, multiyear conversations with him has significantly influenced not only my focus on panentheism and emergence, but also the particular form that these theories have taken in my writing. My debt is deep and lifelong, as is my gratitude. The critical concerns that surface in the following pages do not diminish but rather grow out of that relationship of influence.

Varieties of Naturalism

As his Essay shows, Peacocke is interested in a "naturalistic Christian faith" and is thus engaged in the project of "naturalizing" Christianity. But to what degree? Upon reflection, one realizes that there are many different ways to "naturalize" a faith tradition. Of course, both the process and the product vary widely depending on which tradition one has in mind; naturalizing Judaism or Hinduism has different requirements from naturalizing Christianity. It would turn out, I predict, that formulating a fully naturalized Hinduism, Buddhism, or Judaism is less difficult than fully naturalizing Christianity.

But even within a single tradition, such as Christianity, the project of naturalizing is not monolithic. One discovers a number of different ways that one might approach the task—an impression strengthened, incidentally, by the various responses contained in this volume. Consider these four approaches to naturalizing Christianity. First, one might view Christianity as a whole, its truth claims as well as its normative beliefs, as being of only historical or aesthetic interest, somewhat in the way that stories about the Greek gods still intrigue us today. Second, one might treat Christianity as still having ethical (or moral or political) interest, even though its system of beliefs is false. Although in this case one would not expect people to believe its specific truth claims, one would nonetheless maintain that the Christian tradition expresses certain values that continue to be crucial for living well and for rightly ordering human interactions. Within this group some affirm that other religions express these values in an equally effective fashion, whereas others insist that the Christian story represents a particularly powerful expression of these fundamental values.

A third kind of Christian naturalist retains some truth claims from her tradition, but only those that are fully consistent with science. Miracle claims have to go, she notes, along with all claims for a more-than-natural status for Jesus Christ. In fact, once one has begun to de-supernaturalize, she insists, the project must be carried out to the bitter end, on pain of inconsistency. Thus for example, if "God" means a supernatural entity, the Creator and Sustainer of the universe, then language about God must either be eliminated altogether or thoroughly naturalized, such that the term now refers only to natural features and functions.

Modern theology is chock full of methods for naturalizing theistic language in this way. One thinks immediately of Ludwig Feuerbach, who understood the real referent of theological language to be the unlimited potential of human "species being." Or perhaps language about God might be preserved for the sake of its useful functions, as long as one does not thereby intend actually to refer to any existing divine being or beings. In whichever version, the functionalist strategy has been common within the social sciences from their founding figures until the present, from Emile Durkheim through Karl Marx through Sigmund Freud and on to Pascal Boyer and Scott Atran today.

On this third view, "God" might also be interpreted in the Kantian sense as an idea that is indispensable for practical reasoning, without thereby entailing the claim that a divine being actually exists. (Kantians reinterpret the notions of freedom and immortality in a similar fashion.) Neo-Kantians in the nineteenth century further extended the strategy of treating God as a "regulative" concept: one may speak "as if" there were a god yet without committing oneself to the assertion that some specific being of this type actually exists. The same move is often made with regard to divine action, of course; many use the language of divine action without claiming that any God-initiated events actually occur. All such language may be symbolic, for example—a way of describing how *we* view events in the world, rather than an account of any actual causal influence.

Fourth and finally, the "naturalizer" might preserve language of a Ground or Source of all things yet without asserting that this Ground or Source makes any direct interventions into the natural order. This fourth distinct approach to naturalizing Christianity seems to come closest to what Peacocke has in mind, and in the following pages I will examine it in some detail. It's important to note at the outset that this fourth category actually serves as the heading for at least two rather distinct methods, which we might label the "no influence" and the "no intervention" views. The "no influence" view, usually called *deism*, argues that the Ground or Source cannot have any causal impact at all on the natural world. Advocates include Baruch Spinoza and present-day "Ground of being" theists. The "no intervention" view, which Peacocke holds along with many other scholars in our field, allows in principle for an influence of God on the world, while insisting that the influence is exercised without breaking or setting aside natural laws.

As far as I can tell, no understanding of divine action that is stronger than the "no intervention" view would still count as a naturalized Christianity in any serious sense. Assume, for example, that God intervenes in the natural order in ways that set aside the natural patterns which we call the laws of nature. In this case the resulting outcomes will have to be explained, at least in part, by reference to this supernatural agent, God, and what God has done, *rather than* through the context of natural causes alone. It is the appositive "rather than" which, I take it, the naturalizing project seeks to overcome.

In short, the phrase "naturalistic Christian faith" serves as the rubric for at least four radically different programs, running from Christian atheism on the one side to personal theism with a real (though non-interventionist) sense of divine action on the other. Recognizing the broad spectrum of *degrees* of naturalizing, and the fact that Peacocke stands at the milder end of the continuum of naturalisms, helps one to understand the proposal that he has advanced here. More radical programs of naturalizing allow God-language, if at all, only in the domains of what David Hume famously called the "Before" and "After." Proponents of more radical naturalization do in fact sometimes say that "it's as if" the entire natural order were surrounded by, upheld by, or dependent on "God," but they passionately reject any real assertions of divine influence and often insist on the purely metaphorical (hence, non-metaphysical) character of all God-language.[3] By contrast, as long as one continues to speak of "special divine action," as Peacocke does—that is, as long as one imagines an influence of God on some particular natural event E, such that the full explanation of E must include some reference to that divine activity—one has accepted a partially rather than fully naturalized Christian faith.

The Challenge of Divine Action

We've now located Peacocke's proposal with reference to other projects that might also fall under the rubric of a "naturalistic Christian faith." The task now is to evaluate it on its own terms.

The goal of Peacocke's project is to identify and defend a middle space between two positions on divine action which he believes are untenable: traditional miracle claims on the one hand, and the denial of all special divine action on the other. Thomas Aquinas provided what has become the classic definition of miracles. He distinguished three "degrees" of miracles, noting that "the highest degree in miracles" involves "those works wherein something is done by God that nature can never do. For instance, that two bodies occupy the same place, that the sun recede or stand still, that the sea be divided and make way to passers by. . . . The greater the work done by God, and the further it is removed from the capability of nature, the greater the miracle."[4] Some 700 years later C. S. Lewis defended another version of third-degree miracles in his famous book on divine action: "Nature (at any rate the surface of our own planet) is perforated or pock-marked all over by little orifices at each of which something of a different kind from herself—namely reason—can do things to her." He added, "If God annihilates or creates or deflects a unit of matter He has created a new situation. . . . Immediately all Nature domiciles this new situation, makes it at home in her realm, adapts all other events to it."[5]

This is the sort of divine action that Peacocke's naturalizing project seeks to dispense with, and he studiously avoids affirming any specific miraculous events of this kind. At the same time, he resists the conclusion that *no* special divine action occurs. Contrast his view with that of Maurice Wiles, who defends one of the more radically naturalized versions of Christianity. Wiles retains the notion of God but clearly refuses to make God the explanation of any particular event—which is presumably why advocates of special divine action have so consistently directed their fire in his direction. Wiles does not believe that God carries out intentional acts in the world; hence he denies that "we can properly speak of God being more creative in one place than in another."[6] Yet he still wishes to speak of "the living God, the source of all life and the source of the authentic life which his worshipers seek to realize in grateful awareness of his all-pervasive and sustaining presence."[7] That is, the only act that God actually performs is the universal act of creating and sustaining the world. Humans are nevertheless free to *see* individual events *as* acts of God, and thus to speak of special divine action in this (regulative, symbolic, or fictive) sense. The reason for this strategy is clear. As Gordon Kaufman argues elsewhere, contemporary science presupposes that the natural world is a tightly interconnected web of events; each part of the web is a causal consequence of other parts (perhaps with the addition of some randomness), rather than a self-initiating agent of its own. Kaufman therefore concludes that particular divine actions are "not merely improbable or difficult to believe: they are literally inconceivable."[8]

Clearly Peacocke does not wish to argue for a complete absence of intentional or "special" divine influence, as Wiles and Kaufman do. It turns out, however, that the middle domain between absence and intervention is rather difficult to specify. Many theologians have the intuition that some such *tertium quid* must exist, but it has turned out to be difficult to specify it in a conceptually rigorous manner. Think, for example, of the mediating proposal developed by James Kellenberger, who writes:

Natural miracles occur through God's agency; they are not instances of God's direct action. There is no intervention by God, but God, as creator, is deemed thinkable for establishing the ground of natural events.[9]

As Nicholas Saunders notes:

There appears to be no real distinction on this account between a "natural" event and one of God's actions. . . . By again shifting the emphasis away from causal questions Kellenberger feels able to assert that someone who identifies and believes that a particular event is a natural miracle will be thankful to God for its existence, while if the event is seen as only natural this element of thankfulness will be missing.[10]

Saunders' challenge is clear enough. Yet Peacocke has seen, as Saunders did not, that developing an adequate account of *human* agency in the world—one that is deeply naturalistic while still preserving the distinctive features of personal causality—may be an indispensable step toward formulating a credible theory of divine agency in the world. No theory of human agency is by itself sufficient to prove divine action, of course. But for those who hope to offer a "theology for a scientific age," the detour through human agency may well be necessary.

The Problem of Personal Agency

In a variety of publications Peacocke has described the scientific picture of the world as involving a network of embedded systems. Parts constitute wholes, which themselves become parts within greater wholes, and so forth, until one reaches the entire cosmos as a single interrelated natural system. The scientific evidence in favor of this account is strong enough that it needs no further defense here.

Let's assume that Peacocke has met the first objective of a naturalistic Christian theology, namely to show the compatibility between *at least some form of* Christian faith and this particular scientific picture of the world. A science of embedded causal networks does not need to negate or undercut belief in the existence of God, divine creation, divine sustenance of the world, or divine omnipresence in the strongest possible sense. That God may have had intentions in creating the world, and may continue to have intentions in sustaining it, and that God can be present both to the whole and to every one of its parts, are both perfectly acceptable within this framework. Given these two premises, it is also unproblematic to speak of *the creation of the world as a whole* as a divine act.

So far so good. But Peacocke also wants to say more. In *God and Science* he adds:

> In thus speaking of God's interaction with the world, it has not been possible to avoid speaking of God's single "intentions," of God having purposes, thereby using the language of personal agency. For these ideas of whole-part constraint by God cannot be expounded without relating them to the concept of God as, in some sense, an agent, least misleadingly described as personal.[11]

Surely this is right: what distinguishes theists from deists (and others) is their continuing desire to speak of God's intentions and purposes in the world after the moment of creation. It therefore becomes urgent for us to ask: does Peacocke's framework allow him to speak in this fashion? As we will see, there is at least some reason to worry that it does not.

The core notion that Peacocke uses to describe relations among the different embedded systems or levels of reality from a scientific perspective is *whole-part constraint*. Wholes formed out of parts within nature in turn constrain the behaviors of those parts. (Note his emphasis elsewhere that "'whole-part' is synonymous with 'system-constituent.'"[12]) When Peacocke turns to the questions of personal agency and the God-world relation, he begins by appealing to this notion. What is worrisome about this strategy, however, is that whole-part constraint is not a form of agency, at least not in anything like the normal, intuitive sense of this term. Wholes constrain the behavior of their parts in a passive sense, whereas personal agents are actors who are the active authors of their own actions. Two chemicals may react differently when they are held together by a test tube than they would if the molecules were floating freely in a large body of liquid, but we would not normally say that the test tube is an "agent" in the chemical process. The architecture of the motherboard within your computer constrains the electronic processes that occur, but your motherboard's architecture as such is not an agent. Imagine that it turns out (an assumption I take to be false) that all there is to human thought are the neuronal signals and the electrochemical interactions at the synaptic junctures. Clearly the aggregate that we call the brain as a whole—that is, the sum total of neurons and neural connections for a given individual—constrains the outcome of the neural processes within it. But we would not therefore conclude that the brain as a whole is an agent; rather, we would say that the structural features of the brain affect the behavior of its parts in certain ways. In short, whole-part constraint is not sufficient for agential or intentional language.

Sometimes Peacocke seems to feel the force of this problem. Over the course of his career he has vacillated between speaking of "whole-part constraint" and "top-down causation." Where he speaks of top-down causation, he invariably offers a conceptual framework robust enough to support mental causation and personal agency. At other times—disappointingly—he either treats the two different concepts as interchangeable or seeks to subsume the latter under the former. Thus he writes, for example, "Hence the term *whole-part influence* will be used to represent the net effect of *all* those ways in which a system-as-a-whole, operating from its 'higher' level, is a determining factor in what happens to its constituent parts, the 'lower' level."[13] Now recall Peacocke's view that "whole-part" is synonymous with "system-constituent." Also, let's omit the phrase "operating from its higher level" for the moment, since it's not yet clear in what sense wholes are "operators." Doing so yields the thesis statement, "Hence the term *system-constituent influence* will be used to represent the net effect of all those ways in which a system-as-a-whole . . . is a determining factor in what happens to its constituent parts, the 'lower' level."

But systems as "determining factors" in this sense are not of intentional agents; hence such language is not sufficient to give a (non-reductive) account of intentional agency. Recall the example of the brain and its neurons above. Peacocke seems to recognize the difficulty when he proceeds to assert, "One should perhaps better speak of '*determinative influences*' rather than of 'causation.'"[14] For here's the problem: doesn't the generic term "determinative influence" obscure a crucial distinction—the distinction between the types of "influence" that are sufficient for intentional agency and causality, and the types that are not? If this *is* the crucial distinction, as I think, what are the conditions that must be fulfilled for an instance of "determinative influence" to count as intentional agency?

Three Approaches to Personal Agency

In two recent articles, both appearing in collections from Oxford University Press, Peacocke offers a sophisticated response to this problem, which represents one of the most intriguing proposals available in our field today. The first article, already cited above, offers a brilliant analysis of three different approaches for conceiving the notion of mental causation. The first approach—"Levels H are states of the brain; levels L are individual neuronal events"[15]—is clearly meant to exclude mental causation as such; the causation to which it refers must be neuronal, and perhaps ultimately microphysical.

The second approach at least does not dismiss mental causation altogether: "Levels H are mental-with-brain states; levels L are individual neuronal events."[16] As Peacocke comments (I omit his references to his diagrams), "This is to postulate that the higher-level now mental-with-brain states have a determinative influence, *jointly* with the lower-level neural states, on the succession of mental-with-brain states. . . ."[17] This approach seeks to make room for the insight that the level of the mental is genuinely emergent; it is not reducible to the neuronal or microphysical level. After all, if mental phenomena *are* non-reducible, they must have some effect (assuming, as we must, that "to be real is to do something").

Unfortunately, however, this second approach is not yet able to explain what *is* this more-than-physical causation that produces these new kinds of effects. It must clearly involve more than the "two levels of description" approach that John Searle advocated in 1984, since Peacocke himself criticizes that view.[18] Yet if we are told merely that two dimensions or levels conjointly bring about some effect, we have not yet been given an actual theory of mental causation—apart from the initial claim that something called "the mental" plays *some* role here. Surely, when faced with such an elliptical claim, the principle of parsimony will suffice to direct one's attention toward the type of causality that is better understood—the efficient causality of physical forces. In order to take the "both-and" in this position seriously, one would need to know exactly what it is that "the mental" is supposed to be doing. But those who advocate the second approach do not yet take on this task.

Thus one must turn to Peacocke's third interpretation of mental causation for an account of intentional agency. This account holds that "Levels H are mental states; levels L are brain states," and "mental activity—the content of our consciousness describable in first-person language—is a real emergent *from* brain activity."[19] For "this mental emergence is a distinctive reality which has its own determinant efficacy."[20] Peacocke brings powerful arguments in defense of this third option; together they provide strong support for the conclusion that intentional agency requires mental causal activity. If one rejects mental causation in this third sense, one should conclude that language of intentional agency is illusory, since the kind of causation that it requires does not occur. Only if first-person mental activity is treated as a real emergent, "Could [it] be causally effective on successive brain states . . . Mental events, such as intentions—whatever they are ontologically—have determinative ('causal') efficacy in the physical world. . . . "[21]

Divine Agency and the *Imago Dei* Correlation

Let us use the term *imago dei correlation* to convey the formal connection that almost inevitably exists between one's understanding of God and humanity.[22] The correlation may exist between views of human personhood and divine personhood, or between human and divine agency, or between God's relation to the world on the one hand and the relation of an individual's thought and consciousness to her body on the other. For panentheists, for example, it takes the form of what I have called "the pan-entheistic analogy."[23]

It is thus not surprising that each of Peacocke's three approaches to human consciousness and mentality would produce a separate understanding of the God-world relation and divine action. The first approach does not exclude the existence of God, but it makes any direct influence of God-as-agent on human thought impossible, since it rules out mental or spiritual causation.[24] Nonetheless, as we saw above, there might still be symbolic and figurative uses of Christian language that could be built on top of this ontological platform, as it were. One thinks of the sometimes rich uses of Christian language that one finds among leading deists through the centuries.

The same is true for the second approach. I believe that all the Christological and sacramental language that Peacocke employs in his Essay, above, and most of what occurs in his other publications, *could* still be retained under this interpretation. Although talk of divine effects and mental causes could no longer be given a direct or literal interpretation, symbolic re-interpretations might take up the slack. On this approach the constraining effects of the world-as-a-whole could not literally represent an intentional guidance by God; at least the model provides one with no grounds for making such a claim. Nonetheless, one could always reply, "I picture God to myself as something like an intentional agent who is able to exercise mental (or perhaps better, spiritual) agency. And my model allows me to say that 'the universe as a whole' constrains all of its parts. Thus I shall speak of this highest whole-part constraint *as if* it were the expression of an underlying divine intention." One would have to admit that the language of direct intentional causation is not actually supported by the model, but it's not *inconsistent* with the model either. Advocates of this second view claim that a naturalistic theory of human persons warrants nothing stronger than whole-part constraint, and then, perhaps for religious reasons, they supplement that conclusion with what can only be understood as metaphorical theological language of divine intentional agency at the level of the universe-as-a-whole.

Of course, once one has chosen to define the divine-human relation in this way, one could extend a similar status to much of traditional Christian language. Having assumed that "God" is intending whatever effects follow from universal whole-part constraint, one might then naturally speak of those effects as an influx of divinely-intended information into the system. Since whole-part constraints can in some way influence every part within the system, one could imagine this divine influence as extending also to every individual person. This move might open the door to yet further extensions of theological metaphors. For example, one could imagine that the (divinely intended) informational content from the universe-as-a-whole also applies to oneself, *treating* it *as* a personal communication from *deus pro nobis*, "God for us." Christological and sacramental language could then be added as further metaphorical extensions of the content of this "divine communication." At the same time, by emphasizing that one has drawn one's model from the natural sciences, one might well claim that one's talk of divinely intended content is consistent with the scientific worldview.

Such a use of theological metaphors may not be explicitly ruled out by contemporary science in the way that strong miracles language is. The trouble, however, is that, even if the language of divine agency is not strictly speaking contradicted by science, it is utterly unsupported by any analog in the natural world. Earlier we saw that whole-part constraint by the brain, in the sense of Peacocke's second approach, is not sufficient to count as intentional agency. On what grounds, then, could whole-part constraint justify one in treating the universe-as-a-whole as exhibiting intentional personal agency? The apparent arbitrariness of this move should lead one to give marked preference to the final of Peacocke's three approaches to personal agency—or else to abandon talk of divine action in any sense stronger than what Wiles and Kaufman have advocated.

Personal Divine Agency

Although the third approach implicitly underlies virtually all of Peacocke's theological treatments of the God-world relation, it is worked out explicitly in the two recent papers mentioned above.[25] Peacocke clearly understands God to be a constraining influence on all that exists. It seems obvious that the Ground of all things would be related to the-world-as-a-whole *at least* as strongly as the way in which a system is related to its constituents. But Peacocke decisively supplements this minimal condition by adding the framework of panentheism, that is, the view that the world is contained within the divine, although God is also more than the world. Panentheism offers a way to personalize the divine "whole-part constraint" without falling into pantheism, i.e., the complete identification of God and world.[26]

Peacocke recognizes that his theory of whole-part influence "depend[s] on an analogy only with complex natural systems in general and on the way whole-part influence operates in them."[27] Yet, as we have seen, that particular analogy can't do all the work in the case of the God-world relation, at least not if theism is also to involve the notion of divine personal agency. Thus Peacocke adds, "There is little doubt that [my model] needs to be rendered more cogent by the recognition that, among natural systems, the instance *par excellence* of whole-part influence in a complex system is that of personal agency" (ibid.). Or, as he now writes in the Essay, the God-humanity interaction "evidences *a new kind of causality* of a whole-part kind" (above, 50).

The burning question is whether talk of this new kind of causality, the causality of personal agency, can be justified. On the third approach to mental causation given above, it clearly is; under the first two approaches, I have argued, it is not. If this argument is sound, Peacocke's options become very clear. Only if he is willing to endorse mental causation in the third, stronger sense, as I also am—and assuming that our arguments in defense of the third option hold up—could he be warranted in speaking of divine personal influence on the world. (Again, remember that we are discussing a necessary but not sufficient condition for a theory of divine action.) Only if mental causation is viable[28] can one make sense of theological statements of the sorts he makes in the Essay:

> for when God so acts in a way that can be denoted as an expression of divine grace, then there are effects on human beings that are unique and distinctive, necessitating the variety of classical descriptions of the modalities of grace that we have noted above (see 50).

Divine influence of this sort cannot be merely an instance of whole-part constraint. Rather, it manifests *distinctively personal causation on God's part, causation that makes a difference within the world.*

Whole-part constraint probably suffices for the "ground-of-being theism" that, for example, Wesley Wildman defends. I have argued, however, that as long as Peacocke wishes to defend some form of personal theism, he is obligated to supplement whole-part constraint with a theory of personal causation. One must first defend some form of mental causation: "Persons as such experience themselves as *inter alia* determinative agents with respect to their own bodies and the surrounding world (including other persons), so that the exercise of personal *agency* by individuals transpires to be a paradigm case and supreme exemplar of whole-part influence."[29] Only then can one extend the analogy to argue that

> God could cause particular events and patterns of events to occur which express God's intentions. These would then be the result of "divine action," as distinct from the divine holding in existence of all-that-is, and would not otherwise have happened had God not so intended.[30]

Conclusion

We have found no inherent impossibility to the "naturalistic Christian faith" that Peacocke espouses in this book and elsewhere. Now conservatives sometimes argue that there is no way to preserve Christian language without a "higher" view of divine action. But if their claim is that there is no coherent, consistent way to use Christian language without (say) physical miracles, the claim is not true: Arthur Peacocke's overall program, at least in the form defended here, offers one such way. (Indeed, assuming that even more radical forms of naturalized Christianity than Peacocke's are possible, there may be multiple consistent models. This would mean that internally consistent reinterpretations of the Christian tradition can be found across the spectrum from fundamentalism to Christian atheism—which is not to say that all of these reinterpretations are equally plausible or desirable).

In particular, it turns out to be possible to use Christian theological language with some level of coherence on *any* of the three approaches to mental causation that Peacocke outlines. One need not doubt that a deep religious and devotional attitude, serious moral commitment, and transformative religious experience can occur within each of the three models. The task is therefore not merely to determine whether Christian language *can* be used consistently within a vastly more naturalized context than the worldviews that dominated during most of the history of Christian thought (although Peacocke sometimes writes as if that were the main task facing a "theology for a scientific age"). The harder task is to evaluate what is the best overall balance of naturalism and theism—a burning question for our day that I have only begun to address here.

We did begin the process of evaluation, however. Among the myriad methods for evaluating theological proposals we focused on the issue of coherence, and specifically on the quest for a deeper coherence between one's view of personal agency and one's theory of divine agency, or what I called the "*imago Dei* correlation." Of the three theories of mental causation that Peacocke analyzes, we found that only the third—"mental events, such as intentions . . . have determinative ('causal') efficacy in the physical world"[31]—could do justice to the notion of divine personal agency. It is indeed true that personal agency represents the "paradigm case and supreme exemplar of whole-part influence."[32] This third type of approach must therefore undergird any adequate theory of agency, both in the case of personal agents and in the case of divine agency.[33]

Reflections on the Responses

Arthur Peacocke

These concluding reflections are very much in the spirit of the subtitle of my piece "An Essay in Interpretation," for this book has been intended more as a *ballon d'essai* than as a coherently structured "program"—much as I would like to have lived up to the aspiration of composing a broader and more coherent program in science and theology and of completing it in these pages. Thus it is not surprising that the respondents to the Essay have picked up on very different themes according their own preoccupations and priorities. Still, I have been glad to note that all have valued the exercise as worth attempting in the light of the prevailing culture and my implicit rejection of any form of *super-naturalism* combined with the acceptance of some degree of naturalism.

Because of the inherent variety of the responses, my own reflections and replies have themselves varied in scope and style, consequent upon my own knowledge—or, more often, lack of knowledge—of the particular context within which the respondent is operating (most notably in my response to Donald Braxton). I offer the succeeding remarks in a spirit of promoting further development of these theses and in the hope that such work will prove both possible and fruitful in the future.

Reply to Philip Hefner

I am grateful to Philip Hefner for identifying my project, which I have deliberately and tentatively subtitled an "Essay," as one that accepts "twin givens: the contemporary scientific understanding of the world *and* the classic Christian faith" (above, 60), including belief and liturgical/sacramental worship. Although, as I have argued elsewhere,[1] the content of my Christian belief is (or *should* be) arrived at by inference to the best explanation *(IBE)*, he regards it as "not a minimal vision." I warmly welcome Hefner's perceptive discernment of my intentions in this Essay—standing, as he does, in the rigorous tradition of Lutheran systematic theology. His encouragement in this exercise is all the more welcome in view of the present situation in the science-and-theology dialogue today, in which "theology seems reluctant to integrate the naturalist dimension into its methods and contents, while science is frequently suspicious of transcendence" (60).

As he points out, there is an irony in this situation, since a naturalistic stance in theology is, in fact, closer to the historic "mainstream of Christian faith and theology" (61). In its own times and places theology would have taken as naturalistic a stance as possible to be consistent with the "dogma, which affirms that Christ is 'human, like unto us, divine like unto God,'" thereby establishing "the principle that the entire created order can be a fit vessel for God's presence. Since human nature includes all of nature—physics, chemistry, biology, and psychology—it, too, is capable of receiving the presence of God" (61).

What has happened as the tradition has developed is that natural science has gradually modified our ideas of nature in such a way that traditional formulations of the Christian faith appear, as Hefner rightly argues, to be *anti*natural or inherently *super*natural; consequently, they increasingly go against the grain of a culture that is today deeply indebted to science for its *Weltanschauung*. Of course, this is not the first time that Christianity has interacted deeply with its surrounding culture, for classical Christian theology itself grew out of the Hellenistic thought world. As Hefner notes: "The magnitude of the effort expended in these discussions [involving a Hellenistic framework] underscores the impasse that renders this theological tradition conceptually unusable for us today" (62). There is no need to encumber the expression of Christian belief with that Hellenistic framework, and we are now free to exploit more naturalistic and anti-supernaturalistic notions.

Consistently with all this, Hefner reminds us that the traditional theological axiom, *gratia praesupponit naturam, non destruit sed conservat et perficit eam,* can be translated as *"grace* presupposes nature; it does not destroy it, but rather conserves and perfects it" or "grace *undergirds* nature"(62).

So it would appear that the project of my Essay, in Hefner's eyes, is doing what Christian theology has always done by asking similar questions. But why should such an enterprise appear so radical today? Is it because a "traditional" or "conservative" stance misconceives the task in which theology has been engaged? Or is it just as plausible that science has perhaps raised the stakes by publicly validating inference to the best explanation *(IBE)* as the appropriate way of proceeding to warranted belief in most spheres of intellectual inquiry—thereby questioning the credentials of theology as never before?

In particular, we should proceed without relying on inadequate Hellenistic ideas; we need to "construct concepts of nature and religion that allow compatibility between the two" (64). It is our *idea(s)* of nature that will determine how we conceive of the relation of God to nature, of theology to science (remember R.G. Collingwood). It is in this context that I regard my emphasis on an *emergentist monist* understanding of nature as relevant. It is indeed not yet true that "there is a single authoritative understanding of what it means to take nature seriously, of what constitutes naturalistic philosophy" (64). Still, the recent resurgence of the notion of *emergence*, the wide scope and validity of its applicability to the natural world, encourages one to employ it in this context—and even more so when theological reflection incorporates it into a richer "Emergentist–Naturalistic–Panentheistic" perspective. Although *emergence* as a general description of the nature of nature "is a theory that is yet to be worked out in the full sense" (67), still the recent upsurge of literature on this theme[2] encourages one "to take such a risk." I have taken this risk in my Essay, which I am glad to see that Hefner regards as leaving "no essential element of classic Christian faith out of consideration" (67). Of course, in the end readers must judge for themselves.

A problem, a hazard, remains: would a scientific naturalist recognize the emergent, "spiritual" realities that I claim are to be found in {God + nature + humanity}, and would he/she experience them as *Christian* realities? It is the increasingly urgent task of Christian apologetics to address this question.

Reply to Willem Drees

I am very happy, both by temperament and on reflection, to be regarded as one who is a radical to be conservative. Of course, the hard question is often: How much of each? I suspect Willem Drees and I allocate their respective weights differently; certainly we do so in different contexts of the science-religion discussion. His response is so methodical that it merits an equally methodical and, I hope, rational reaction from myself.

Drees refers positively to earlier statements of mine to the effect that if matter is said to give "rise to humans, this does not downgrade humans. Our esteem for matter should be modified. . . . Matter has the potential to become a Wolfgang Amadeus Mozart, a Gautama Buddha, and a Jesus" (quoted above, 71). He takes this as agreed by all the denizens of the anti-reductionist camp—even though there are many finer distinctions to be drawn and the case has to be made again in each different context. In his first two points Drees himself shows how the concept of "emergence," as the other side of the coin to "reduction," is helpful in establishing a naturalistic Christian faith, not least in seeing divine design as present not in separate phenomena but in the basics of the whole structure.

I am sorry that he had the impression that I think God is "found more easily in higher-level phenomena than in lower-level ones" (above, 73). I meant rather that entities that display higher-level phenomena are more likely to have a richer relation to God, even if we cannot fully analyze that relation. One cannot discount the long history of the divine Presence as being experienced personally by individuals, which would seem at least to establish that *this* higher, emergent level, the level of persons, is the appropriate source of metaphors for divine agency—for human persons are indeed *agents* in the natural world.

As regards the process philosophy of David Ray Griffin, to which I refer in discussing forms of naturalism, my position is that, insofar as Griffin postulates God as entering into the natural process at all, then he differs not only from those (e.g., Karl Peters and Charley Hardwick) who only go as far as an existentialist interpretation of "God," but even more so from those pure "naturalists" who eliminate "God" entirely from the relationships of the natural world—an interpretation with which I think Drees concurs (see, e.g., 73 above).

Divine temporality

There is no doubt that the Einsteinian revolution in our conception of time has raised a new issue for theists, insofar as time is now considered as much an integral part of what has been created as matter/energy or space. The panentheistic component incorporated into the "Emergentist–Naturalistic–Panentheistic" (*ENP*) perspective is very relevant here. For if "all-that-is" is in some sense *in* God and is created by God, then we have to think of God as generating time as the matrix of all events that involve matter/energy in space. So I, with some others and because of the Einsteinian revolution, accept a dipolarity in God—a transcending by God of time, which is however then given existence by God "in" God. There is indeed an increased complexity within the divine nature because of this proposed dipolarity, but I believe it is one we cannot avoid. God, as it were, creates time "as God goes along," as the milieu in which all else is located. The existence of time is the necessary condition for there to be a dynamic, creative *process*, since process is of course an inherently temporal concept. Furthermore, the processes of time involve, for many living creatures, various degrees of suffering, so that the concept of God creating their lives *within* Godself makes intelligible the notion that God can suffer "in, with, and under" their trials.

Among those who espouse the idea of panentheism, I have not been one who relies on the analogy that the world is to God as our bodies are to our minds (or to us as "persons"), for the reasons I spell out in the Essay (see Chapter 4). There is a danger, of course, in using any feature of the created world (minds, persons) as a model for the relation of the Creator to what-is-created. So, for example, we have to be careful about speaking of the "mind of

God," even as we affirm that God displays mental activities in God's relationships with us.

Deism

The reason why Aubrey Moore (above, 63 and 18), and others since, including myself, have argued that geological transformation and biological evolution weighed (and still weigh, when one considers the matter of cosmological development) against deism is that these scientific developments show that the creation of new forms—terrestrial, astronomical, and biological—was and still is a *process*. As long as God must be creating through a process in time, we are required to speak of God as acting *all the time* as Creator. Hence a stronger emphasis on the immanence of God in the world is required than hitherto, and certainly more than "deism" accommodates. Indeed, deism seems to discard altogether any idea of God's immanence in the world, and in any case to underplay it. The readers must judge for themselves what other changes in terminology for God's creative Presence in the world are required in light of the "historical sciences." In my view, the *ENP* perspective—incorporating emergentist monism, theistic naturalism, and panentheism—becomes necessary as a result, even though it leads to a more naturalist understanding of the God-world relation. *Pace* Willem Drees, I think modifying deism into "a more theistic" position (above, 76) is inadequate to make the strong emphasis on immanence that is now required.

In my treatment, I intend the independent existence and ultimate transcendence of God to be maintained and preserved through all the alterations in our view of the world that are incumbent upon the results of new knowledge from the sciences.

New Testament studies

On the topic of New Testament history, Drees's own judgment is that all the historical criteria[3] one uses should be those of general historical scholarship. I would want to stress in response that, in the case of the history of Jesus, the data considered by historical scholarship point toward a uniqueness that is inescapable intellectually and compelling personally. Again, fine and different lines will be drawn by different commentators. I agree that many sciences do not allow *repeatable* observations for their testing procedures, yet it is of the essence of *historical* judgment to endure this limitation. As regards the resurrection of Jesus, my own judgment is that the historical evidence for the baffling appearances of the risen Jesus are strong,[4] but those for the empty tomb less so; as a result, the latter are ultimately less significant for faith, and hence also theologically less significant. A subjective element in such historical judgments is inevitable, and I gratefully acknowledge Drees's positive reaction to my "religious affirmations" (79-80, and Essay, Chapter 6).

Reply to Christopher Knight

Christopher Knight not only accepts the "Emergentist–Naturalistic–Panentheistic" (*ENP*) perspective that I urge as a foundation for Christian theology, but he also wants to utilize other building materials than those I employ as "elements of the superstructure" of a new synthesis. He sees a problem in nomenclature with respect to the term "theistic naturalism," and he classifies my form of this proposal as "weak" because "a specific divine initiative" is still called upon, so that it cannot avoid being regarded as "supernaturalist." His "strong theistic naturalism," by contrast, refers to the *absence* of specific divine action in any form and employs no concept of divine response.

Knight points to a parallel ambiguity in my use of the term "panentheism," finding an ambiguity in the way that I envisage the causal joint between God as agent and the world upon which God acts as agent, specifically in the question of whether it is external or internal to God. This, I take it, is his distinction between a panentheism that is "weak" (allowing the laws of nature to be autonomous and so outside God) and one that is "strong" (such that the laws of nature are not entirely autonomous and can, indeed have to, be manipulated by God to effect his will).

Is this a confusion on my part, or a *necessary* tension that inevitably arises when one is propounding the "sacramentality of the created order" (above, 84)—a tension that arises also when interpreting the eucharist and the incarnation and which my Essay is intended to resolve or, at least, to reduce? I am intrigued by the point in Knight's response at which he argues that it *is* possible to manifest a *strong* theistic naturalism and a *strong* panentheism in a non-deistic framework in the light of a "neo-Byzantine model of God's presence and action in the world" (above, 84ff). I acknowledge the inadequacy of my use of this strand of Christian theology to refute what has, apparently, appeared to some to be my susceptibility to deism (a position I do *not* wish to adopt, as Knight notes), and I welcome his enriching of our resources for understanding God's action in the world by his eastward turn. I would hope that my bringing together in one metaphysical framework (see Chapter 5 above) how to understand both the operation of grace and the operation of natural processes can now be seen as an explication of Knight's statement that "the Eastern Christian tradition knows nothing of . . . 'pure nature,'" since it sees grace as being "implied in the act of creation itself." And, because of this, the cosmos is seen as inherently "dynamic . . . [,] tending always to its final end."

I would hope also that my exposition in the light of the *ENP* perspective might allow more than a "few (outside of the Eastern Orthodox tradition)" (above, 85) to find it a useful "philosophical articulation of this [Eastern] model."

Knight then proceeds to explore the "teleological-Christological" character of the vision he is articulating. I welcome this inasmuch as I find it a fruitful attempt to circumvent various dilemmas with which we would otherwise be confronted—in particular, that posed by the tendency of a strong theistic naturalism to precipitate into an eighteenth-century deism. He goes so far as to urge that "a strong theistic naturalism can at least in principle be constructed in such a way that the scope of divine action is not limited in the way the deists assumed. Because of this, the chief theological objection to a strong theistic naturalism is rendered void" (above, 88). My (and others') "weak" panentheism could then draw the sting of the criticism that deism implies an absentee-landlord-God. Knight's neo-Byzantine (as he calls it) "cosmic vision" also allows him, he argues, to cope with the question of teleology. If he is right in this contention, we can only be grateful for his helping us to break out of our overly Western-orientated perceptions. But is it a successful ploy?—Only the future will tell.

Reply to Karl Peters

Karl Peters and I share a sense of the imperative to understand the world naturalistically, in the sense that we both prefer to give weight to scientific accounts of unusual phenomena—both in the natural and in the human context—rather than to import any supernatural agency to explain them. ("Miracles," in the breaking-of-laws sense, are very improbable.) So we both eschew God as a *deus ex machina* operating in the natural and/or human worlds. He therefore avoids, in this context, the use of all God-language in which "God" refers to a distinct ontological entity, whereas I want to draw on such language as necessary for a complete interpretation of "spiritual"[5] experiences.

Hence I am intrigued by the way Peters finds himself drawing distinctions between himself and me concerning the God-world relation in a way that, on closer examination, seems to dissolve his position into mine. This is particularly obvious in his challenging, and moving, account of his and his companions' experience of a eucharist in a ritualistic setting to which they were not accustomed and which he was drawn to interpret personalistically—"drawn," it appears, and to use his own phrase, "toward theological personalism." More philosophically, when he is distinguishing between the different kinds of transcendence we individually experience, he rightly sees us both as rejecting supernaturalism *and also* a "substantive" view of God. But if by this last phrase he means God seen, in some regard, as "at least personal,"[6] then I fear he may have obscured the differences between us.

For I wish to emphasize that, in experiences in which those undergoing them distinguish not only humanity and nature but also "God,"[7] this last term refs to the effect of a distinctive reality on those other two components in the experience. In the "complex"[8] of God–nature–humanity there emerges the effect of new realities, the least obvious of which, *a priori*, is "God." "God" is often discerned in these situations as "at least personal"—as an ontologically transcendent entity, because new realities are discerned in experiencing the complex. The pressure to do so comes, in my case, not from "the authority of the Christian tradition" as such, but, rather, because the language of that tradition has rightly taken account of the need to recognize the historical (and in the experience of Christians, contemporary) existence of new distinctive realities as "determinative causes," which we then need new concepts to describe. This is parallel to the need I have found to employ the word "person" to designate what is distinctive of the mind-brain-body complex of interactions, and to the use of my *third* interpretation[9] of the possible relations of higher to lower levels to explicate this complex.

It appears that, *pace* Karl Peters's own proposal, my attempt to expound a naturalistic interpretation of spiritual experiences cannot avoid utilizing ontologically transcendent and personalistic terminology for the divine "component" in complex spiritual experiences. Indeed, Peters' own moving account of his experience at that eucharist seems to me to indicate the fruitfulness of such terminology in "transcending" (*mot juste!*) such differences as remain between us.

Reply to Donald Braxton

Life occasionally has its moments of serendipidity[10] that transpire to have a significance far beyond their first appearance. Such was the case when, long after completing my Essay, the then-current copy of *Zygon* landed on my desk containing an article by Donald Braxton on "Naturalizing Transactions in the New Cosmologies of Emergence."[11] I realized at once that he was moving in the same thought-world as myself as we probe ways of expressing the Christian faith devoid of supernaturalistic elements. The problem in such an exploration is that, in eliminating *super*naturalism, one may not do justice to the potential transcendence-within-immanence character of the natural world to which the "Emergentist–Naturalistic–Panentheistic" (*ENP*) perspective points. As I have argued, this perspective enables one to recast Christian theology in a fruitful new way consistent with the contemporary understanding of emergence that science presses upon us.

Donald Braxton was, I am glad to say, persuaded to respond to my Essay. He has done so by first working through the significance of *sacrament* and *sacrifice* in the communities of *Homo sapiens* at various stages of their evolution. I am not competent to adjudicate on his interpretations. If they are correct—and I have no reason for thinking otherwise—then he successfully shows that "sacrament and sacrifice are in effect the cultic and cultural experiences of the processes regulating the emergence and maintenance of religious communities"(above, 105). This is a general judgment from an anthropological viewpoint, of course, but it can also be translated into the context of the Christian tradition: "the Christian community is the in-gathered body of Christ transformed in the death and resurrection, and sustained by the abiding presence of the Holy Spirit" (above, 105).

Braxton believes—and this is the crucial point—that "these dimensions of Christian community can be rendered intelligible on completely naturalistic terms." Furthermore, "Cast against the backdrop of this evolutionary history, Christian communities become once again 'available' to those spiritually seeking individuals held at a distance by their disinclination to embrace supernaturalism" (above, 105). His argument, based on the evolutionary history of religion among *Homo sapiens,* leads him into a naturalistic interpretation of sacrament and sacrifice in the Christian tradition. When one also incorporates ethnographic work on the culture of science, this view turns out to involve experiential correlates between sacrament and sacrifice, forming something like analogical bridges between the worlds of altar and laboratory and between religious community and scientific cooperation.

Braxton finds that my *ENP* perspective "opens up new pathways to understanding the ancient religious practices of sacrament and sacrifice . . . in a manner that eschews supernaturalism yet does justice to the experiential power of sacrament in religious community." He thus offers "a naturalistic interpretation of sacramentality consilient with modern scientific theory, especially the current fascination with emergence" (above, 111). Readers must judge his argument for themselves. But I was particularly struck by the passage:

> That story [of a meaningful world in which our understandings of who we are can be fused with what we must do in response to our situation] is best told in our day and age through our common evolutionary heritage as a species among species. *Our* presence is "miraculous" not because it is an exception to the way the world is, but *because* this is the way the world is, fecund with, and generative of, emergent properties. . . . To be Christian in the scientifically informed age is to see the Christian story as a chapter in this longer and larger narrative. In a word, it is about vision and the behaviors that result from a world well seen. (above, 112-13, first emphasis added)

My Essay may be regarded as an attempt to advance the *ENP* proposal as the basis of such a vision. As I am grateful to Donald Braxton for giving it this context, I have deliberately chosen in this response to highlight core themes in his.

Reply to Ann Pederson

Ann Pederson responds positively to the broadly "sacramental panentheism" of my Essay; like Christopher Knight, she also points eastward for theological resources in her emphasis on *theosis*, the divinization of humanity as the "work of Christ." I am, on the whole, happy with this designation, although the significance of the suffering of Christ must never be underplayed in the great Christian scheme—Christ is Pantocrator *because* he is also Christ as Savior, and he can redeem only *because* he is Pantocrator.

Her response reveals the deep and, to me as an Anglican, previously unacknowledged resources of traditional Lutheran theology in sharing my "sacramentalist" approach—especially in its slogan of *finitum capax infiniti,* which I interpret as an imperative for us all to become "co-creating creatures."[12]

Even more chastening is her pointing to the recent contributions of feminist writers—in particular the work of Donna Haraway, a feminist philosopher of science—as helping in unexpected ways to dissolve the boundaries between what had previously been conceived of as *pure* categories, *inter alia*: the human and the non-human, organic (carbon) and technological (silicon), freedom and structure, history and myth, diversity and depletion, modernity and postmodernity (above, 126). If not taken too far, this project would be consistent with my sacramentalist approach, which emphasizes positively the natural as a vehicle of the divine—"in, with, and under" the natural—while maintaining, in my case,[13] their necessary distinction.

Reply to Nancey Murphy

At one point in her response Nancey Murphy draws attention to my long espousal of "critical realism" (above, 131), which I have adopted as the most viable epistemology for both science and theology. This, I have always accepted, has kept me firmly in the "modern" camp.[14] For example, I have found science to be a bulwark against those forms of postmodernism (which turn out to be most of them!) that undermine any realist reference, certainly within science itself. The success of science in prediction and control seemed to me, as it does to most scientists, a sufficient validation of the reality of that to which scientific terms refer. I am therefore intrigued to find myself, in Nancey Murphy's response, cast, as it were, in the role of the sheriff riding out into the western sun to rescue the captives held in the wagons of modernism!

According to her, I achieve this, however, not so much by refuting the postmodernist claim that "the laws of nature are only known as they are *fabricated* in the laboratory by the new *social order* of scientists" (above, 131)—though I have in the past tried to do just that—but rather by allowing "talk about God," that is, theology, a coherent and proper place, *pace* the deliverances of modernism, in the hierarchy of knowledge of nature, humanity, and society. For, according to modernism, with respect to science "no one is truly modern who does not agree to keep God from interfering with Natural Law as well as with the laws of the Republic" (above, 131, quoting Bruno Latour). She regards my placing of theology at the *top* of the hierarchy of the "sciences" (that is, the intellectual disciplines) as helping to reunite God, Nature, and Society in a way to which modernism could never aspire.

Does all this make me a *Postmodern* Prophet, as her title indicates? I am surprised to be so designated (and perhaps not a little gratified!). Yet I must admit to some discomfort since, by placing theology at the pinnacle of the hierarchy of knowledge of complexity, I thought I was allowing *more* realistic references to God, nature, and humanity (society) than other ways of locating theology would allow—if in fact they allow any.

However, whatever the labels (modernism, postmodernism, or post-post-modernism), Nancey Murphy and I seem to agree about the epistemological landscape and the reality of reference of terms in different disciplines—though she, I think, would continue to give more weight than I do to their social conditioning and thus social modification.

A substantial part of her response is an acute analysis of the change in attitude regarding the autonomy of morals that arose in modern thought, consequent upon the supposed absence in modernity of any deity providing an ultimate purpose or *telos* to human life. My locating theology in the space at the top of the hierarchy of the sciences that describe human behavior, a space left vacant by modernity, provides (as I understand her) a way of escaping from this *impasse*—of filling this vacuum that modernism has, historically, created. Readers must here, as elsewhere, judge for themselves. It is true that I have sought to show that there are forms of epistemology that *can* face up to the major crisis facing Christianity in the modern era that is constituted by the rise of empirical science. The latter part of Murphy's response makes it clear that my thesis could be useful not only for "harmonizing theology with the scientific worldview" (above, 139) but also for qualifying that autonomy of morals to which modernism gives rise.

Reply to Robert Russell

I cannot but be grateful to Bob Russell for his comprehensive and sympathetic listing, in his first section, of various facets of the science-and-religion dialogue on which I have expressed my views—sometimes "orthodox," sometimes less so. He approves of the metaphors that I use to depict how the "critical realist" view can account for the way that *both* science and theology reach their conclusions as "candidates for reality," that is, as truly referring to reality in their respective propositions, however tentatively and revisably. Or, rather, as I still am (alas) impelled to urge: how theology *ought* to reach its conclusions in such a manner as to (attempt to) match the warrant that science itself can command. These standards often lead me—for example, in Trinitarian doctrine and eschatology—to positions further removed from traditional Christian formulations than Russell himself. He is always generous in his serious consideration of my positions, and I sometimes wonder why we move in different directions theologically. Is it because I started my scientific life more as an experimentalist than he did and so give more weight to historical considerations (e.g., biblical criticism) than to more intellectual considerations such as coherence? Perhaps.

Russell finds my presentation of the "hierarchy of the disciplines" to be fruitful in relation to the science and theology dialogue,[15] particularly when combined with Ian Barbour's way of relating scientific and theological methodology. It helps to generate in his note 25 an "eightfold way" of "creative mutual interaction," a very useful model within which other investigators will, I am sure, be eager to locate their own explorations.

It is in the ways we view God's action in the world that we have had our principal differences. Russell, too, wishes to exploit the notion of top-down causation[16]; he agrees that the unpredictability of chaotic systems is not a useful crevice into which to fit God's action in the world. But he then turns to quantum theory as his resource, a move I have found uncongenial in the past for the reasons he accurately reports. I have concluded that, from a quantum perspective, in which the future is only probabilistically predictable, God could know it only to that extent. Bob Russell takes the more classical—and theologically traditional—Boethian[17] view (above, 146–47) that God knows the future in its present actuality, a view that we might now, in the light of relativistic physics, call the four-dimensional "block view" of the universe. This view is supported by many physicist-theologians and would render doubtful any claim that God does not know the future directly. Along with a weighty number of others, I regard it as still an open question, so the debate on this question continues.

Russell goes on in his response to make a major contribution to the wider debate when he deals with the question of what kind of "interface" God has with the world. Because of my emphasis on God acting on the "world-as-a-whole," I have been at pains to assert that the interface where God acts in a top-down causative fashion must be "everywhere." We both agree that, in any theological perspective, this boundary is ontological; in any Christian formulation of the relation of Creator to creation there is, and always will be, an ontological distinction between what the world is and what God is. What Bob Russell significantly points our attention to, however, is that, according to current scientific cosmology, *the universe does not have a boundary*, that is, a physical boundary. Thus he attempts to take us out of a threatening cul-de-sac by invoking "divine action at every point-like quantum event throughout the universe" (above, 151)—what he calls "top-down-through-bottom-up"—thereby rescuing us all from this dilemma. Perhaps I shall, after all, have to swallow my aversion to involving quantum events as the locus of "special divine action" that is not intervention in divine top-down causation!

Reply to Keith Ward

Keith Ward's response was, I found, one of the most difficult to reply to—for, somehow, he contrives by a subtle legerdemain to claim me not only as a dualist concerning human nature, but indeed to translate my careful emergentist monism, incorporated in the *ENP* perspective, into the more traditional language of "persons." As a result of the process of transformation, he can argue that we both concur entirely.

He affirms, for example, that "human persons are interfused complexes of spirit and matter but they have in their nature both spirit and matter,"[18] thereby emphasizing the category of spirit and a dualism of spirit-matter. In this definition of human person, "spirit" represents a distinct entity and thus implies (to complete his quotation) "a basic dualism that is founded on the absolute priority of the pure spirit of God." Yet I have always argued that the only dualism I find acceptable is that between God ("Spirit," if you like) and the world ("nature," including human nature). I have always regarded human beings as psychosomatic unities, and this is where the concept of "emergence" is so useful and clarifying. For its use in this context prevents one from designating human nature (or any other constituent of the world) in a dualistic fashion, regarded as a kind of mixture of distinct ontological entities. Rather, each "complex" displays emergent properties dependent on its constituents. "Person"[19] comes to mean that distinctive mutual organization of the whole to which the word "person" is referring.

Similar considerations also explain my apparent bias against "miracles"[20]—or, to be more precise, my requiring especially strong historical evidence for them (which is often not forthcoming)—on the grounds that we are sufficiently aware of the *regular* behavior of many particular natural "complexes" to preclude disruption of their behavior. Talk of the possibility of a new "regime" operating in the occurrence of miracles, in which unusual events are deemed specifically to express a particular intention of God is, in my view, part of the problem of "special divine action," which has been intensively examined in the conferences and volumes supported jointly by the Vatican Observatory and the Center for the Theology and Natural Sciences in Berkeley. (It was in that context that I first drew in the notion of whole/part or top/down causation.) No doubt the idea that in certain complexes there emerge new properties could also be brought into consideration in relation to what are described as "miracles." My own view of "laws of nature," which become relevant to this context, is that they are *observed* regularities; hence the interpretation of any disruption of such regularities has to be carefully weighed in each reported case. For example, speaking of "the miracle of the eucharist," as some do, is best interpreted, in my view, within the sort of framework employed in my Essay.

I think Keith Ward and I are both striving for similar objectives—I from an "Emergentist–Naturalistic–Panentheistic" perspective, and he from a perspective that employs the notion of "person" regarded as a fusion of the *distinct* entities of "spirit" and "matter." I find the *ENP* perspective more consistent with the scientific worldview in that it recognizes the gradations in emergent properties that we actually observe in the world. This entails a much stronger emphasis on the immanence of God in the world and on God's omnipresence.

When I read Keith Ward's response, I feel, "Yes, that's *something like* what I want to say." Still, I continue to think my way of placing it within an *ENP* perspective has more traction with scientific understandings of the all-pervasive regularities of the world, which the sciences go on unveiling to an astonishing degree. But his shrewd response merits a much fuller and more detailed analysis than is possible here.

Reply to Philip Clayton

By distinguishing between a variety of ways of "naturalizing" Christianity (above, 164-66), Philip Clayton has been able to clarify for me the "program" that my Essay attempts to execute. He is right in placing my position at "the most minimal end of the spectrum" of naturalization, which runs from the most fully possible naturalized faith at one end (a Christian atheism) to a "personal theism with a real (though non-interventionist) sense of divine action" (166) at the other. He claims that, "along with many other scholars in our field" (166)—presumably, for example, the participants in the CTNS/VO program of conferences[21]—I espouse a "no intervention" view that "allows in principle for an influence of God on the world, while insisting that the influence is exercised without breaking or setting aside natural laws" (166).

I think this represents my position fairly, capturing the spirit of the exploration in my Essay of the possibility of "a partially rather than fully naturalized Christian faith" (166). I am also indebted to Philip Clayton for another clarification in his response, namely, of the relation between two strands in my thinking—the one represented by the previous paragraph on a "Christian naturalism," and the other concerned with postulating human agents as "personal," including the question of whether or not this connection could serve as a suitable model for divine agency in the natural world. This latter project itself involves two vexed questions: the notorious question of the mind-brain-body relation, and the question of divine agency, namely is it "personal" or otherwise? Clayton points out that I am seeking "to identify and defend a middle space between two positions on divine action" that I find to be untenable: "traditional miracle claims on the one hand, and the denial of all special action on the other" (166).

It is encouraging that, in the end, he does not think this objective is unattainable as long as I keep to my *third* interpretation of mental causation,[22] in which "mental emergence is a distinctive reality which has its own determinant efficacy" (171), since this distinctive reality is part of the strong support for intentional agency requiring mental causal activity.

It is not an Editor's usual duty to coordinate and unify the dispersed elements in an author's work. I am grateful to Philip Clayton for making it clear how a positive relation between the different strands of my thinking can provide a sound and coherent metaphysical basis for a minimalized form of the Christian faith.

Nunc Dimittis

Up until July 2004 I was blessed with a long, healthy and fruitful life. In July 2004, in my eightieth year I was diagnosed not only with prostate cancer, but having it in an advanced form. This was an enormous shock to myself and my wife who was with me in all the medical consultations.

The hormone therapy which I was prescribed was of the simplest form and limited my public activities very little. However when I visited St Petersburg in the spring of 2005 to attend a conference I was finding walking very difficult, even for very short distances. Consequently I had to cancel two holidays, many conferences and lectures which would have taken me abroad. It was only during this time that the enormity of what I had to face up to gradually dawned on me and this catalysed me to finishing off "An essay in interpretation" concerned with a more naturalistic understanding of the Christian faith which I hoped would be congenial to more orthodox believers as well as those who are seriously challenged by the scientific world view as the norm for their thinking.

I was also happy to see the fruition of my co-operation with Ann Pederson in *The Music of Creation*, which was published in November 2005 with an accompanying CD to give musical illustrations of the text.

Meanwhile I was much less mobile but not so much that I could not spend Christmas with my daughter in 2005. By this time I was taking an enormous range of pills, bouts of nausea were becoming frequent, and it was becoming less and less possible to envisage a normal life of any kind.

I was trying to be stoic and trying not to inveigh against God for what was clearly going to be my fate—a fate I had not really envisaged or imagined.

However I did manage, with the help of my wife and daughter, shuffling me between cars, hotels and view points to visit for a few days my beloved Cairngorms and Strathspey, but clearly life was rapidly changing. And not only in this regard.

It became clear—during a short stay in hospital where I was treated (unsuccessfully) for excessive swelling in my legs—that the house we had lived in for a long time would be impossible to manage with one person an invalid. Hence the house was sold and we bought a small flat that would be suitable for the two of us. The move, especially clearing out the accumulation of over twenty years would have been impossible without the energetic and willing help of my grandson David who gave up his summer vacation to do this.

I had only a week living in the flat before the cancer struck again, totally immobilising me to the extent that I was taken to Sir Michael Sobell House. This is a wonderfully caring hospice that brought me through near fatal kidney failure. I was cared for there for five weeks and then moved to a nursing home from which I am at present writing and where one day is very like another. I experience discomfort as I am washed and hoisted by carers, for I am paralysed from the waist downwards. This alternates with relatively pain-free periods when I can read, listen to music and enjoy the company of friends and family who faithfully and regularly visit me.

I have long been one of those who have been unsure about the role and efficacy of intercessory prayer. My view of it was that the intercessor by placing him or herself in the presence of God, with the person prayed for very much in mind, enabled that person to experience the enfolding presence of God. I felt that the person prayed for was being taken up in the loving arms of God enhancing the divine presence. I can honestly say that this is what I have experienced. Many many people from the science-religion community, a wide circle of friends and of course my family have assured me that they are praying for me. It seems that my suffering has evoked a response from friends and colleagues which has revealed to me (surprisingly) how my words and actions have been a positive influence in their lives. This kind of prayerful support I had not expected; it was, and is a great help.

Uniquely through all of this the mutual love of my wife and myself has been enriched and deepened in her daily visits and the knowledge that we share the same prayers and the conviction that death will not part us.

Over the years I have given much thought and spilt much ink on the nature of God and God's interaction with people. Not surprisingly the subtler nuances of my deliberations have fallen away before the absolute conviction that God is love and eternally so. This remains the foundation of my prayers and thoughts for "underneath are the everlasting arms." This is not always easily experienced and it needs much concentrated meditation—the "black dog" of depression is sometimes difficult to expel.

Another of my concerns over the years has been the recurrence of what theologians call "natural evil." I have often attempted to illustrate the ambivalence of this concept, for example showing that what we call natural evil is a consequence of a divinely created lawlike structure implementing the divine purpose to bring into existence intelligent persons. The irony is that one of the examples I took was the role of mutations in DNA which are the basic source of evolution, and so of the emergence of human beings—and also of cancer. This is a new challenge to the integrity of my past thinking. I am only enabled to meet this challenge by my root conviction that God is Love as revealed supremely in the life, death and resurrection of Jesus the Christ.

However the fact remains that death for me is imminent and of this I have no fear because of that belief. This conviction was not available to the non-Christian audience who, according to Bede, were addressed concerning the mystery of life.

> *"Such", he said, "O King, seems to me the present life of men on earth, in comparison with that time which to us is uncertain, as if when on a winter's night you sit feasting with your ealdormen and theigns,— a single sparrow should fly swiftly into the hall, and coming in at one door, instantly fly out through another. In that time in which it is indoors it is indeed not touched by the fury of winter, but yet, this smallest space of calmness being passed almost in a flash, from winter going into winter again, it is lost to your eyes. Somewhat like this appears the life of man; but of what follows or what went before, we are utterly ignorant."*

Thanks to the revelation of God through Jesus the Christ we do not share this ignorance. I know that God is waiting for me to be enfolded in love.

Death comes to every one and this is my time.

Arthur Peacocke
2006

Appendix A: "God"

The best explanation of all-that-is and all-that-is-becoming is an:

Ultimate Reality, "God,"
who

- is the self-existent Ground of Being, giving existence to and sustaining in existence all-that-is;
- is One—but a diversity-in-unity, a Being of unfathomable richness;
- includes and penetrates all-that-is, but whose Being is more than and is not exhausted by it (panentheism);
- is supremely and unsurpassedly rational;
- is the immanent Creator, creating in and through the processes of the natural order;
- is omniscient, with only a probabilistic knowledge of the outcomes of some events;
- is omnipotent, self-limited by God's nature as Love;
- gives existence to each segment of time for all-that-is-becoming;
- is omnipresent to all past and present events and will be to all future ones;
- is eternal, exists at all times—past, present, and future;
- transcends past and present created time (but does not know the future, since it does not exist to know);
- possesses a dipolarity in relation to time, being transcendent but also experiencing succession in relation to events and persons; hence
- is not "timeless," but is temporally, and so personally, related to humanity;
- is (at least) personal, or supra-personal—yet also has impersonal features;
- is the ultimate ground and source of both law ("necessity") and "chance"—an Improvisor of unsurpassed ingenuity;
- has something akin to "joy" and "delight" in creation;
- suffers in, with, and under the creative processes of the world;
- took a risk in creation;
- is an Agent who holistically affects the state of the world-System and thereby, mediated by whole-part influences, can affect particular patterns of events to express divine intentions;
- communicates with human persons through the constituents of the world (in religious and other experiences) by imparting meaning and significance to particular patterns of events.

From Arthur Peacocke, *Paths from Science towards God* (Oxford: One World, 2001), 129-30.

Appendix B: A Short Life of Jesus

Extracts from an appendix to *The Historical Jesus: A Comprehensive Guide* by Gerd Theissen and Annette Merz (SCM Press, London,1998), 569-72.

Jesus was born in Nazareth shortly before the end of the reign of Herod I (37-4 BCE), the son of Joseph, a craftsman in wood and stone, and his wife Mary. He had several brothers and sisters. We know the names of some of his brothers. He must have had an elementary Jewish education, was familiar with the great religious traditions of his people, taught in synagogues and was called "rabbi" during his public activity.

In the 20s of the first century CE he joined the movement of John the Baptist, who was calling on all Israelites to repent and promising them deliverance in the imminent judgment of God by a baptism in the Jordan. Here John offered the forgiveness of sins in ritual form—independently of the possibilities of the temple in providing atonement. This was a vote of no confidence in the central religious institution of Judaism, which had become ineffective. Jesus, too, had himself baptized by John. Like everyone else he confessed his sins. Like everyone else he, too, expected the imminent judgment of God.

Soon Jesus made an appearance independently of John the Baptist—with a related message, but one which put more emphasis on the grace of God that still gives everyone a chance and allows more time. . . . Jesus' basic certainty was in fact that a final shift in the direction of the good had taken place. Satan had been conquered, and in essentials evil had been overcome. One could experience this in exorcisms, in which the demons had to flee.

With this message Jesus travelled through Palestine as a homeless itinerant preacher, focussing his attention on small places north-west of the Sea of Galilee. He chose twelve disciples from among ordinary people, fishermen and farmers with Peter at their head. They were representatives of the twelve tribes of Israel with whom he wanted to "rule" the Israel that would soon be restored. His idea was a kind of "representative popular rule". . . .

At the centre of Jesus' message stood Jewish belief in God: for Jesus, God was a tremendous ethical energy which would soon change the world to bring deliverance to the poor, the weak and the sick. However, it could become the "hell-fire" of judgment for all those who did not allow themselves to be grasped by it. Everyone had a choice. Everyone had a chance, particularly those who by religious standards were failures and losers. Jesus sought fellowship with them, the "toll collectors and sinners." He found prostitutes more open to his message than the pious. He was confident of his power to move people to repentance. He did not call for any demonstration of repentance, nor any baptism. For him God's grace was certain without such rites.

In his picture of God Jesus combined two traditional images in a new way.

For him, God was father and king. However, he never spoke of God as king but always only of God's "kingdom" . . .

The most impressive of his words were the parables, little poetic narratives which ordinary people could understand. However, in them he inculcated an "aristocratic" self-confidence: everyone had infinite responsibility before God, and in view of all that could risk their whole lives. Salvation and damnation were now near.

At the same time Jesus was active as a charismatic leader. People flocked to him in order to profit from his gift of healing. He saw these healings as signs of the kingdom of God which was already beginning, and at the same time as an expression of the power of human faith. . . .

The great transformation of the world by God was also to change human wills: Jesus' ethical teaching was the pattern for a human being who was governed entirely by the divine will. He intensified the universalistic aspects of the Jewish Torah and dealt in a "liberal" way with those ritual aspects which distinguished Jews from Gentiles. But all his teaching remained grounded in the Torah. He put the commandment to love God and neighbour at the centre of his ethic, but he radicalized it so that it became an obligation to love enemies, strangers and the religious outcasts. In ritual questions he was demonstrably non-fundamentalist. . . .

What he taught to all has to be distinguished from the demands he placed on his followers, men and women: here in individual instances he could require transgressions against the Torah. . . . Here he called for a radical ethic of freedom from family, possessions, home and security. . . .

Jesus attracted attention and provoked opposition by his teaching and life. He discussed his behaviour with the Pharisees, precisely because in many things he was close to them. They both wanted the whole of life to be penetrated by the will of God, but argued over the way in which this could be done. . . . It was Jesus' criticism of the temple which was fateful for him when he went up to Jerusalem for the Passover. . . . Jesus attacked it [the temple] directly: he prophesied that God would replace the old temple with a new one. By a symbolic action, the so-called cleansing of the temple, he disrupted the temple cult and deliberately provoked the aristocracy associated with the temple. For his disciples . . . at the last meal that they shared he instituted a new rite: a simple meal which he shared with them one day before the beginning of the Passover in expectation of a dramatic escalation of the conflict with the Jerusalem aristocracy. Probably . . . he hovered between expecting death and the hope that God would intervene before his own death and usher his rule. The aristocracy which arrested him took steps against him because of his criticism of the temple, but accused him before Pilate of a political crime, of having sought power as a royal pretender. In fact many among the people and his followers expected that he would become the royal Messiah who would lead . . . Israel to new power. Jesus did not dissociate himself from this expectation before Pilate. He could not. For he was convinced that this God would bring about the great turning-point in favour of Israel and the world. He was condemned as a political troublemaker and crucified with two bandits (very probably in April 30 CE). His disciples had fled. However, some women disciples were braver, and witnessed the crucifixion from afar.

After his death Jesus appeared first either to Peter or to Mary Magdalene, then to several disciples together. They became convinced that he was alive. Their expectation that God would finally intervene to bring about salvation had been fulfilled differently from the way for which they had hoped. They had to reinterpret Jesus' whole fate and his person. They recognized that he was the Messiah, but he was a suffering Messiah, and that they had not reckoned with. . . . Now they saw that he was "the man" to whom according to a prophecy in Dan. 7 God would give all power in heaven and on earth. For them Jesus took a place alongside God. Christian faith had been born as a variant of Judaism: a messianic Judaism which only gradually separated from its mother religion in the course of the first century. . . .

Appendix C: God's Interaction with the World

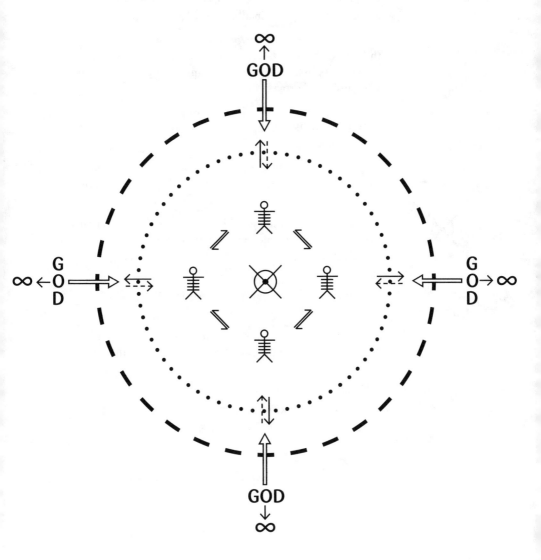

Fig. 1. Diagram representing spatially the ontological relation of, and the interactions between , God and the world (including humanity). See page 200 for legend to this diagram.

Legend for figure 1:

GOD is represented by the whole surface of the page, imagined to extend to infinity (∞) in all directions.

 the WORLD, all-that is created and other than God, and including both humanity and systems of non-human entities, structures, and processes.

 the human WORLD: excluding systems of non-human entities, structures, and processes.

 God's interaction with and influence on the world and its events

 tip and shaft of a similar double-shafted arrow perpendicular to the page: God's influence and activity *within* the world.

 effects of the non-human world on humanity
human agency in the non-human world

 personal interactions, both individual and social, between human beings, including cultural and historical influences

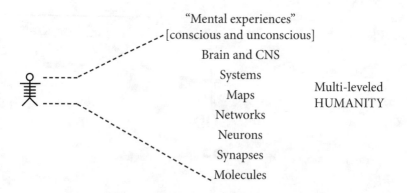

"Mental experiences"
[conscious and unconscious]
Brain and CNS
Systems
Maps — Multi-leveled HUMANITY
Networks
Neurons
Synapses
Molecules

Apart from the top one, these are the levels of organization of the human nervous system depicted in fig. 1 of Patricia S. Churchland and T. J. Sejnowski, "Perspectives on Cognitive Neuroscience," *Science* 242 (1988): 741–45.

Notes

A Naturalistic Christian Faith for the Twentieth Century / Peacocke

1. For example, in my *Creation and the World of Science* (Oxford: Oxford University Press, 1979; 2nd edition, 2004); *Theology for a Scientific Age* (London: SCM Press/Minneapolis: Fortress Press, 1990, 2nd enlarged edition, 1993) (henceforth *TSA*); *God and Science: A Quest for Christian Credibility* (London: SCM Press, 1996); *Paths from Science towards God: The End of All Our Exploring* (Oxford: One World, 2001) (henceforth *PSG*).

2. In "Emergence, Mind and Divine Action: The Hierarchy of the Sciences in Relation to the Human Mind-Brain-Body," in *The Palace of Glory: God's World and Science* (Adelaide: ATF Press, 2005), 91–118; in *The Re-emergence of Emergence*, eds. P. Clayton and P. Davies (Oxford: Oxford University Press, 2006); the preceding article also appears in a slightly modified form in *International J. of Philosophy,* no.1, July, 2004, 33–51. See also my "Emergent Realities with Causal Efficacy—Some Philosophical and Theological Applications," in *Evolution and Emergence: Systems, Organisms, Persons*, eds. N. Murphy and W. R. Stoeger, S.J. (Oxford: Oxford University Press, forthcoming 2007).

3. I use this term, rather than "natural," because I wish to set my interpretation in that contemporary context of discussion which emphasizes that the real world is best understood in the terms that the sciences deliver.

4. For a parallel naturalistic formulation, see, in particular, Charley D. Hardwick in *Events of Grace: Naturalism, Existentialism, and Theology* (Cambridge: Cambridge University Press, 1996).

5. As does C. Hardwick, ibid.

6. Ibid.

7. Ibid., 5ff.

8. Derived from Rem Edwards, *Reason and Dialogue: An Introduction to the Philosophy of Religion* (New York: Harcourt, Brace, Jovanovich, 1972), 133–41.

9. Hardwick, *Events of Grace,* 5–6 (emphasis added).

10. Ibid., 8. In (c), "conservation of value" is Hardwick's broad and loose term to characterize the way theologians discuss a final destiny for humanity beyond death in the forms of both objective and subjective immortality.

11. Ursula Goodenough, *The Sacred Depths of Nature* (New York: Oxford University Press, 1998).

12. As quoted by Britt Peterson in *Science and Theology News* 6/4 (December, 2005): 15.

13. Karl E. Peters, *Dancing with the Sacred: Evolution, Ecology, and God* (Harrisburg, Pa.: Trinity Press International, 2002), 1.

14. Karl E. Peters, "Dancing with the Sacred—Excerpts," *Zygon* 40 (2005): 631.

15. Ibid., 9.

16. Where by "cues" he is referring to "the subtle initiatives a given situation offers . . . the

delicate signs that nature or other people give of how things are flowing," Peters, ibid., 48–49, quoting from Denise and John Carmody, *Christianity: An Introduction* (Belmont, Calif.: Wadsworth, 1983), 20, in which they call the creativity that continually gives rise to new structures, new life forms, new thoughts and practices in society, the "the dance of God."

17. Peters, *Dance of the Sacred,* 49.

18. David R.Griffin, "Scientific Naturalism: A Great Truth That Got Distorted," *Theology and Science* 2 (2004): 11.

19. Ibid., 11.

20. Ibid., 9–30.

21. Ibid., 27.

22. Summarized in Appendix A with respect to its exposition of the term "God," taken from *PSG,* 129–30.

23. All quotations in this sentence and the following are from Hardwick's Preface in *Events of Grace*, xi.

24. V. Lossky, *The Mystical Theology of the Eastern Church* (Cambridge: James Clarke, 1991), 70.

25. Thomas Traherne, *Centuries,* 1670; First Century, vol. 18 (London: The Faith Press, 1960, 1963), 9.

26. Acts 17:28 (AV and RSV).

27. Dante, *Paradiso,* trans. G. L. Bickersteth (Oxford: Shakespeare Head Press, Blackwell, 1965), Canto I, ll: 1–3, 517.

28. *Bhagavadgita*, translation of S. Radhakrishnan, *The Bhagavadgita* (Bombay: Blackie & Son, 1971 [1948]), Chap. VI (30), 204.

29. Ibid.

30. See, for example, Arthur Peacocke, *Intimations of Reality: Critical Realism in Science and Theology* (Notre Dame, Ind.: University of Notre Dame Press, 1984).

31. The account of these developments in Part One given here follows closely the expositions in my "Emergent Realities with Causal Efficacy—Some Philosophical and Theological Applications," in Murphy and Stoeger, eds., *Evolution and Emergence*; and for Chapters 3 and 4, especially, see my *PSG*, Chap. 8.

32. Conventionally said to run from the "lower," less complex, to "higher," more complex systems, from parts to wholes, so that these wholes themselves constitute parts of more complex entities—rather like a series of Russian stack dolls. In the complex systems I have in mind here, the parts retain their identity and properties as isolated individual entities.

33. See, e.g., Arthur Peacocke, *TSA,* 36–43, 214–18, and figure 1 (based on a scheme of W. Bechtel and A. Abrahamson, in their figure 8.1 in *Connectionism and the Mind* (Oxford and Cambridge, Mass.: Blackwell, 1991).

34. W. C. Wimsatt has elaborated these criteria of "robustness" for such attributions of reality to emergent properties at the higher levels in his "Robustness, Reliability and Multiple-Determination in Science," in *Knowing and Validating in the Social Sciences: A Tribute to Donald T.Campbell,* eds. M.Brewer and B.Collins (San Francisco: Jossey-Bass, 1981).

35. William Bechtel and Robert C. Richardson, "Emergent Phenomena and Complex Systems," in *Emergence or Reductionism?* eds. A. Beckermann, H. Flohr, and J. Kim (Berlin and New York: de Gruyter, 1992), 266, emphasis added (page references in the text are to this article).

36. As does Philip Clayton, in "Neuroscience, the Person and God: An Emergentist Account," in *Neuroscience and the Person: Scientific Perspectives on Divine Action,* eds. R. J. Russell, N. Murphy, T. C. Meyering, & M. A. Arbib (Vatican City State and Berkeley, Calif.: Vatican Observatory and the Center for Theology and the Natural Sciences, 1999), 209; and in his *Mind and Emergence* (Oxford: Oxford University Press, 2004), 60.

37. D. T. Campbell, "'Downward Causation' in Hierarchically Organised Systems," in *Studies in the Philosophy of Biology: Reduction and Related Problems,* eds. F. J. Ayala and T. Dobhzhansky (London: Macmillan, 1974), 179–86.

38. For a survey with references, see Arthur R. Peacocke, *The Physical Chemistry of Biological Organization* (Oxford: Clarendon Press, 1983, 1989).

39. Harold Morowitz, *The Emergence of Everything* (New York: Oxford University Press, 2002).

40. I. Prigogine and I. Stengers, *Order Out of Chaos* (London: Heinemann, 1984).

41. Terence Deacon, "Three Levels of Emergent Phenomena," paper presented to the *Science and Spiritual Quest Boston Conference*, October 21–23, 2001. Similar proposals are made by him in "The Hierarchic Logic of Emergence: Untangling the Interdependence of Evolution and Self-Organization," in *Evolution and Learning: The Baldwin Effect Reconsidered*, eds. B. Weber & D. Depew (Cambridge, Mass.: MIT Press, 2003). See also B. Weber and T. Deacon, *Cybernetics & Human Knowing* 7 (2000): 21–43.

42. This whole-part influence often also extends to the external environment of the system in question.

43. J. C. Puddefoot, in "Information and Creation," in *The Science and Theology of Information*, eds. C. Wassermann, R. Kirby, and B. Rordoff (Geneva: University of Geneva, 1991), 7–25, has carefully clarified the relation between the different uses of "information": first, physicists, communication engineers, and neuroscientists use it as referring to the probability of one outcome among many possible outcomes of a situation; second, there is the meaning of "to give shape or form to" (stemming from the Latin *informare*), that is, a pattern-forming influence; finally, there is the ordinary sense of information as knowledge, broadly indicating something like "meaning."

44. A dictum attributed to S. Alexander by J. Kim in "Non-Reductivism and Mental Causation," in *Mental Causation*, eds. J. Heil and A. Mele (Oxford: Clarendon Press, 1993), 204; and in his "'Downward Causation' in Emergentism and Nonreductive Physicalism," in Beckermann, Flohr, Kim, eds., *Emergence or Reduction?* 134–35.

45. *A New Dictionary of Christian Theology*, eds. Alan Richardson and John Bowden (London: SCM Press, 1983).

46. *Shorter Oxford English Dictionary*, 1973.

47. Richardson and Bowden, *A New Dictionary*, 148. As the dictionary goes on to point out: "The common use of 'theism,' in contrast, does not have these negative implications."

48. A. Moore, "The Christian Doctrine of God," in *Lux Mundi*, 12th edition, ed. C. Gore (London: Murray, 1891), 73.

49. H. Drummond, *The Lowell Lectures on the Ascent of Man* (London: Hodder and Stoughton, 1894), 428.

50. C. Kingsley, *The Water Babies* (London: Hodder and Stoughton, 1930 [1863]), 248.

51. H. van Till, "The Creation: Intelligently Designed or Optimally Equipped?" *Theology Today* 55 (1998): 349, 351.

52. The capacity for the form of music to serve as a model for divine creativity has recently been expounded and illustrated (musically) in *The Music of Creation* (with CD) by Arthur Peacocke and Ann Pederson (Minneapolis: Fortress Press, 2005); and the capacity of dance in my *Creation and the World of Science* (Oxford: Oxford University Press, 1979, 2nd edition, 2004), 106ff.

53. Michael W. Brierley, "Naming a Quiet Revolution: The Panentheistic Turn in Modern Theology," in *In Whom We Live and Move and Have Our Being: Panentheistic Reflections on God's Presence in a Scientific World*, eds. Philip Clayton and Arthur Peacocke (Grand Rapids, Mich.: Eeerdmans, 2004), 1–15. The whole volume constitutes a survey of the various understandings (and misunderstandings) of the concept of panentheism.

54. According to the definition given in the *Oxford Dictionary of the Christian Church* (Oxford: Oxford University Press, 1985), 1027.

55. Augustine, *Confessions,* VII, 7.

56. *Cf.* the musical analogy regarding Beethoven, above.

57. For an elaboration of this thesis, see my "Emergence, Mind and Divine Action: The Hierarchy of the Sciences in Relation to the Human Body-Mind-Brain." Philip Clayton, in his *Mind*

and Emergence, argues that consciousness is one example (among others) of what he denotes as "strong emergence."

58. See n. 43, above.

59. According to Bishop Kallistos of Diokleia in "Through the Creation to the Creator," *Eco-theology* 2 (1997): 12–14.

60. It cannot be too strongly emphasized that "causal" does not here, nor elsewhere in this paper, refer to a Humean succession of cause/effect events but rather to the influence of the state of the whole of a complex on the properties and behavior of its constituent parts.

61. And of the external environment, when this is appropriate.

62. *The Myth of God Incarnate*, ed. John Hick (London: SCM Press, 1977).

63. John A. T. Robinson, *The Human Face of God* (London: SCM Press, 1973), 68 and n.3.

64. 2 Cor. 5:19.

65. For example, Geza Vermes, *Jesus the Jew* (London: SCM Press, 2001); E. P. Sanders, *Jesus and Judaism* (London: SCM Press; Minneapolis: Fortress Press, 1985).

66. See, for example, Arthur Peacocke "DNA of our DNA," in *The Birth of Jesus*, ed. George J. Brooke (Edinburgh: T. and T. Clark, 2000), 59–67.

67. Raymond E. Brown, *The Birth of the Messiah* (London: Chapman, 1977), 527, emphasis in text.

68. John Macquarrie, *Jesus Christ in Modern Thought* (London: SCM Press, 1990; Philadelphia: Trinity Press International, 1990), 392–93.

69. Gregory of Nazianus, *E 101*, quoted by H. Bettenson, ed. *Documents of the Christian Church* (London: Oxford University Press, 1943 and 1956), 64.

70. Sanders, *Jesus and Judaism*, 11.

71. But not to Jews, it must be noted.

72. Gerd Theissen and Annette Merz, *The Historical Jesus: A Comprehensive Guide* (London: SCM Press; Minneapolis: Fortress Press, 1998), 569–72. For extracts from this text, see Appendix B to this volume.

73. Following Van A. Harvey, *The Historian and the Believer: The Morality of Historical Knowledge and Christian Belief* (New York: Macmillan, 1966).

74. Quotations from Harvey, ibid., 194–96, made by Hardwick in his *Events of Grace* (n. 4 above), 234–35, whose exposition I follow closely here.

75. Harvey, *The Historian and the Believer*, 270, as quoted by Hardwick, *Events of Grace*, 235.

76. Christopher Rowland, *Christian Origins* (London: SPCK, 1985), 244.

77. See, for example, my *TSA*, 269–79.

78. We have already discussed one of these, the virginal conception of Jesus by Mary; see *TSA*, 274–88.

79. Pheme Perkins, *Resurrection: The New Testament Witness and Contemporary Reflection* (London: Chapman, 1984).

80. See *TSA*, 285–6.

81. For further elaboration of, and support for, these ideas, see: Christopher C. Knight, "Hysteria, Myth: the Psychology of the Resurrection Appearances," in *Modern Churchman* 31:2 (1989): 38ff.; and *idem*, "Resurrection, Religion and 'Mere' Psychology," *International Journal for the Philosophy of Religion* 39 (1996), 159ff.

82. As expounded in *TSA*, Chap. 11.

83. C. F. Evans, "Resurrection," in Richardson and Bowden, *A New Dictionary*.

84. Even if one does not affirm the virginal conception (a "supernatural" event), for in a theistic-naturalistic *ENP* perspective God is present in all natural events and, in some, is especially revelatory of Godself.

85. The experience of the disciples of the risen Christ is crucial here, whether or not the tomb was empty. They encountered him in his undisputable identity in a form that appeared to them as bodily and knew he had been taken through death and was fully present in his complete personhood to his Father (an interpretation I have elaborated in *TSA*, 279ff.).

86. Corresponding to Van Harvey's level (iv), Chap. 6 above, concerning the meaning of "Jesus of Nazareth."

87. James D. G. Dunn, *Christology in the Making* (London: SCM Press, 1980), 262 (emphasis omitted).

88. A theme developed in Arthur Peacocke, "The Incarnation of the Informing Self-Expressive Word of God," in *Religion and Science: History, Method, Dialogue*, eds. W. Mark Richardson and Wesley J. Wildman (New York and London: Routledge, 1996), 321–39.

89. John Bowker, *Religious Imagination and the Sense of God* (Oxford: Clarendon Press, 1978), 187–88.

90. John 1:1-18, esp. vv. 1-5.

91. Macquarrie, *Jesus Christ in Modern Thought*, 43–44, 106–7, 108.

92. Cf. Prov. 8:22-31.

93. Furthermore, in Hellenistic Judaism, especially in Philo Judaeus (*ca.* 30 B.C.E.–45 C.E.), *Logos* was "a cosmological principle equated with the image or work of God but distinct from God himself and so intermediate between God and the world" (C. B. Kerferd in *The Encyclopedia of Philosophy*, ed. P. Edwards (New York and London: Collier-Macmillan), 5: 84).

94. Macquarrie, *Jesus Christ in Modern Thought*, 392.

95. An interpretation of the eucharist I originally suggested in "Matter in the Theological and Scientific Perspectives," in *Thinking about the Eucharist*, a collection of essays by members of the Doctrine Commission of the Church of England (London: SCM Press, 1972), 28–37 (especially 32); and (with some additions) in my *God and the New Biology* (London: Dent, 1986; reprinted in Gloucester, Mass.: Peter Smith, 1994), chap. 9, 124–25. It is entirely congruent with the view expounded by Niels Henrik Gregersen in "God's Public Traffic: Holist versus Physicalist Supervenience," in *The Human Person and Theology*, eds. N. H. Gregersen, W. B. Drees, and U. Görman (Edinburgh: T. and T. Clark, 2000), 180–82.

96. As quoted by Kathleen Raine, *William Blake* (London: Thames and Hudson, 1970), 187.

97. Gregersen, "God's Public Traffic."

98. "Naturalistic" in the sense I have earlier adopted. For an elaboration of this proposal, see *TSA*, 160–66. For its history and development, see my "God's Interaction with the World: The Implications of Deterministic 'Chaos' and of Interconnected and Interdependent Complexity," in *Chaos and Complexity: Scientific Perspectives on Divine Action*, eds. R. J. Russell, N. Murphy and A. R. Peacocke (Vatican City State and Berkeley, Calif.: Vatican Observatory and The Center for Theology and the Natural Sciences, 1995), 263 n. 1; and also "The Sound of Sheer Silence: How Does God Communicate with Humanity?" in *Neuroscience and the Person: Scientific Perspectives on Divine Action*, eds. R. J. Russell, N. Murphy, T. C. Meyering, and M. A. Arbib (Vatican City State and Berkeley, Calif.: Vatican Observatory and The Center for Theology and the Natural Sciences, 1995), 21, n. 1.

99. *N.B.*, the same may be said of *human* agency in the world. Note also that this proposal recognizes more explicitly than is usually expressed that the so-called "laws" and regularities that constitute the sciences usually apply only to certain perceived, if ill-defined, levels within the complex hierarchies of nature.

100. *TSA*, 161 and 164. John Polkinghorne has made a similar proposal in terms of the divine input of "active information" in his *Scientists as Theologians* (London: SPCK, 1996), 36–37.

101. See n. 43.

102. See the series of volumes sub-titled "Scientific Perspectives on Divine Action" emanating from extended consultations organized by the Center for Theology and the Natural Sciences, Berkeley, and the Vatican Observatory.

103. I would not wish to tie this proposal too tightly to a "flow of information" interpretation of the mind-brain-body problem.

104. And so, implicitly, patterns of human brain activity.

105. E.g., in *TSA*, 112 and *passim*—and eschewing any talk of God as "a person."

106. Church of England, *Common Worship* (London: Church House Publishing, 2000), 457.

107. *Oxford Dictionary of the Christian Church*, eds. F. L. Cross, and E. A. Livingstone (Oxford: Oxford University Press, 1974), 586.

108. Ibid., 587.

109. Ibid. , 447.

110. Ibid., 1320.

111. Ibid., 1121.

112. Some of whom have also identified created, uncreated, extrinsic, elevating, and sanctifying "grace."

113. Cf. the article, "Grace," by E. J. Yarnold in Richardson and Bowden, eds., *A New Dictionary*, 244–45.

114. By, for example, the Religious Experience Research Unit/Centre, Oxford ; also see G. Ahern, *Spiritual/Religious Experience in Modern Society* (Oxford: RERU, 1990) and T. Beardsworth, *A Sense of Presence* (Oxford: RERU, 1977).

115. Gordon D. Kaufman, *God—Mystery—Diversity: Christian Theology in a Pluralistic World* (Minneapolis: Fortress Press, 1996), 101–9. Serendipity is defined as: "The faculty of making happy and unexpected discoveries by accident" (*Oxford English Dictionary*).

116. Ibid., 109.

117. Ps. 103: 13–17; also see *Common Worship, Daily Prayer* of the Church of England (London: Church House, 2005), 800–801.

118. As sacraments are so defined in the Catechism (an addition made to it in 1604) of *The Book of Common Prayer* of the Church of England: "an outward and visible sign of an inward and spiritual grace."

119. With all the qualifications I have noted earlier of the model of the world as "God's body" (Chap. 4, this essay).

120. Chap. 9, this essay.

121. Chap. 8, this essay.

122. A theme more fully developed by Philip Hefner in *The Human Factor: Evolution, Culture, and Religion* (Mineapolis: Fortress Press, 1993).

123. A Lutheran phrase usually employed to describe the presence of God in the sacraments.

124. George Herbert, "Providence," in *Works of George Herbert* (Oxford: Clarendon Press, 1941), 116ff.

125. A theme fully discussed in *The Work of Love: Creation as Kenosis*, ed. J. Polkinghorne (Grand Rapids: Eerdmans, 2001).

126. A plea I have begun to make elsewhere, in "Complexity, Emergence and Divine Creativity," in *From Complexity to Life*, ed. N. H. Gregersen (Oxford: Oxford University Press, 2003), 201–2; and one I have long since adumbrated in my Bampton Lectures of 1978, *Creation and the World of Science* (Oxford: Clarendon Press, 1979, 2d edition, 2004), Appendix C, on "Reductionism and Religion-and-Science: [Theology] the Queen of the Sciences?" 367–71.

Response 1. Arthur Peacocke's Theology of Possibilities / Hefner

1. Numbers within parentheses refer to pages of the opening Essay in this volume by Arthur Peacocke. I assume that the reader is familiar with Peacocke's new essay; hence I will keep exposition of the work to a minimum.

2. Peacocke's Essay, 6.

3. Ibid., 8-9.

4. P. K. Moser and J. D. Trout, eds., *Contemporary Materialism: A Reader* (London: Routledge, 1995).

5. Essay, 17.

6. See the chapter by Ann Pederson in this volume, particularly her emphasis on the finite as bearer of the infinite.

7. Bernhard Stoeckle, *Gratia Supponit Naturam: Geschichte und Analyse eines theologischen Axioms,* Studia Anselmiana philosophica theological, fasc. 49 (Rome: "Orbis Catholicus," Herder, 1962).

8. Essay, 50.

9. Ibid., 14.

10. Arthur Peacocke, *Intimations of Reality: Critical Realism in Science and Religion* (Notre Dame, Ind.: University of Notre Dame Press, 1984), 80.

11. R. G. Collingwood, *The Idea of Nature* (Oxford: The Clarendon Press, 1945).

12. *Religion and Science: Spiritual Quest for Meaning* (Kitchener, Ont.: Pandora Press, in press). My analysis deals with David Griffin, Arthur Peacocke, and Karl Peters.

13. A glance at his scientific treatise, *The Physical Chemistry of Biological Organization* (Oxford: Clarendon Press, 1989[1983]), reveals a great deal about the scientific ideas that have influenced Peacocke's concept of nature and the correlated theological vision.

14. Among these other thinkers, I have in mind Karl Peters, David Griffin, and Charles Hardwick.

15. Essay, 12.

16. Ibid., 12.

17. This "ladder" or hierarchy of wholes is elaborated in detail in Peacocke's *Theology for a Scientific Age,* 2nd enlarged edition (Minneapolis: Fortress Press, 1993), 216–17, figure 3, derived, with modifications, from a scheme of W. Bechtel and A. Abrahamsen in *Connectionism and the Mind* (Oxford and Cambridge, Mass.: Blackwell, 1991), figure 8.1. The same diagram, but with Theology and Religion included in the uppermost level, is to be found in Peacocke's "A Map of Scientific Knowledge, Genetics, Evolution, and Theology," in *Science and Theology: The New Consonance,* ed. Ted Peters (Boulder. Colo. and Oxford: Westview Press, 1998), 189–210; and in his "Relating Genetics to Theology on the Map of Human Knowledge," in *Science and Theology: Controlling Our Destinies: Philosophical, Historical and Religious Visions on the Human Genome Project,* ed., Phillip Sloan (Notre Dame, Ind: University of Notre Dame Press, 1999), 343–66.

18. Essay, 14.

19. Ibid., 15.

20. Ibid., 27.

21. It cannot be too strongly emphasized that "causal" does not here, or elsewhere in this paper, refer to a Humean succession of cause/effect events, but to the influence of the state of the whole of a complex on the properties and behavior of its constituent parts.

22. And of the external environment, when this is appropriate.

23. Essay, 27-28.

24. Ibid., 27.

25. Alfred North Whitehead, *The Concept of Nature* (Ann Arbor, Mich.: University of Michigan, 1957), Chap. 1.

26. Alfred North Whitehead, "The Organization of Thought," in *Limits of Language,* ed. Walker Gibson (New York: Hill and Wang, 1962), 12.

27. Mihaly Csikszentmihalyi, "Consciousness for the Twenty-First Century," *Zygon* 26:1 (March 1991), 17–18.

28. See Ann Pederson's discussion of Sittler's thought in her chapter in this volume.

29. William Temple, *Nature, Man, and God* (London: The Macmillan Co., 1935), 478.

Response 2. Some Words in Favor of Reductionism . . . / Drees

1. Arthur Peacocke, *Paths from Science towards God: The End of all our Exploring* (Oxford: One World, 2001).

2. For my own attempt regarding these distinctions, see Willem B. Drees, *Religion, Science and Naturalism* (Cambridge: Cambridge University Press, 1996); and also Willem B. Drees, "Religious Naturalism and Science," in Philip Clayton, ed., *The Oxford Handbook of Religion and Science* (Oxford: Oxford University Press, 2006) .

3. See for my objections to the process ontology, Drees, *Religion, Science and Naturalism,* 257–59; for Griffin's dislike of my position, see D. R. Griffin, "A Richer or a Poorer Naturalism? A Critique of Willem Drees's *Religion, Science, and Naturalism,*" *Zygon* 32/4 (December 1997), 593–614; and his *Religion and Scientific Naturalism: Overcoming the Conflicts* (Albany: State University of New York Press, 2000).

4. Augustine, *Confessiones,* Book XI. See for my own considerations, Willem B. Drees, "A Case against Temporal Critical Realism? Consequences of Quantum Cosmology for Theology," in Robert J. Russell, Nancey Murphy, C. J. Isham, eds., *Quantum Cosmology and the Laws of Nature: Scientific Perspectives on Divine Action* Vatican City State: Vatican Observatory Publications, and Berkeley, Calif: Center for Theology and the Natural Sciences, 1993); and Willem B. Drees, *Beyond the Big Bang: Quantum Cosmologies and God* (La Salle, Ill: Open Court, 1990), 141–50.

5. An example is Frank J. Tipler, *The Physics of Immortality* (New York: Doubleday, 1994), in which the author reconstructs notions such as immortality and resurrection as possibilities, without addressing what context ideas about immortality or resurrection arose in. The analogy of theological reinterpretation and the development of theories in the natural sciences was developed at greater length in Willem B. Drees, "The Significance of Scientific Images: A Naturalist Stance," in Niels Henrik Gregersen and J. Wentzel van Huyssteen, eds., *Rethinking Theology and Science: Six Models for the Current Dialogue* (Grand Rapids: Eerdmans, 1998), 87–120, 111–12.

6. Peacocke refers in the beginning of his Essay to Charles Hardwick, *Events of Grace: Naturalism, Existentialism, and Theology* (Cambridge: Cambridge University Press, 1996), but does not align himself with Hardwick's existentialist approach.

Response 3. Emergence, Naturalism, and Panentheism / Knight

1. Arthur Peacocke, "A Naturalistic Christian Faith for the Twenty-First Century" (henceforth "NCF"), 20 above.

2. NCF, ch. 2.

3. See Philip Clayton, *God and Contemporary Science* (Edinburgh: Edinburgh University Press, 1997), for a discussion of both.

4. NCF, 20 above.

5. See, e.g., the essays in Philip Clayton and Arthur Peacocke, eds., *In Whom We Live and Move and Have Our Being: Panentheistic Reflections on God's Presence in a Scientific World* (Grand Rapid: Eerdmans, 2004).

6. Christopher C. Knight, *Wrestling with the Divine: Religion, Science, and Revelation* (Minneapolis: Fortress Press, 2001), ch. 2.

7. See, e.g., Arthur R. Peacocke, "God's Interaction with the World: The Implications of Deterministic 'Chaos' and of Interconnected and Interdependent Complexity," in R. J. Russell, N. Murphy and A. R. Peacocke, eds., *Chaos and Complexity: Perspectives on Divine Action* (Vatican City State and Berkeley, Calif.: Vatican Observatory and Center for Theology and the Natural Sciences, 1995), where he comments on how "unhelpful" the "distinction between creation and providence often proves to be" (283).

8. John Polkinghorne, *Scientists as Theologians: A Comparison of the Writings of Ian Barbour, Arthur Peacocke and John Polkinghorne* (London: SPCK, 1996), 26ff.

9. Christopher C. Knight, "Divine Action: A Neo-Byzantine Model," *International Journal for Philosophy of Religion* 58 (2005), 181ff., the arguments of which are expanded in my book in preparation, provisionally entitled, *The God of Nature: Incarnation and Contemporary Science,* Theology and the Sciences (Minneapolis: Fortress Press, 2007).

10. Aspects of the earlier history of this approach are excellently summarized in Eugene TeSelle, "Divine Action: The Doctrinal Tradition," in Brian Hebblethwaite and Edward Henderson, eds., *Divine Action: Studies Inspired by the Philosophical Theology of Austin Farrer* (Edinburgh: T. and T. Clark, 1990), 71ff.

11. See Andrew Louth, "The Cosmic Vision of St. Maximos the Confessor" in Clayton and Peacocke, eds., *In Whom We Live,* 184ff., for a good brief discussion of the use of the term *logos*

in Greek thought. When properly understood, this usage allows and even demands what—when translated into English—might seem little more than a play on words.

12. Stephen W. Need, "Re-Reading the Prologue: Incarnation and Creation in John 1:1-18," *Theology* 106 (2003): 403. This article provides a good brief description of why Western exegesis is now beginning to recognize the biblical roots of the intimate link between creation and the Christ-event.

13. Vladimir Lossky, *The Mystical Theology of the Eastern Church* (Cambridge: James Clarke, 1957), 101.

14. Because Byzantine authors were rarely systematic, it is in fact difficult to say whether they posited a unified approach to divine action in which a single model was sufficient. Although they did not speak of "special" divine action in the technical sense, they often spoke rather loosely in a way that might now be interpreted within that framework. In what follows, however, I advocate a neo-Byzantine model that firmly avoids the concept of "special" action. In this sense, it is inspired by the earlier Byzantine model, but is not simply a restatement of it.

15. Kallistos Ware, Bishop of Diokleia, "God Immanent yet Transcendent: The Divine Energies according to St. Gregory Palamas," in Clayton and Peacocke, eds., *In Whom We Live*, 160.

16. See n. 14.

17. It also, incidentally, indicates that providence that is impersonal in mechanism need not be impersonal in either the giver's intention or the recipient's perception—thus answering the sort of objection voiced, for example, by John Polkinghorne, *Science and Christian Belief: Theological Reflections of a Bottom-Up Thinker* (London: SPCK, 1994), 78ff.

18. Christopher C. Knight, "Naturalism and Faith: Friends or Foes?" *Theology*, 107 (2005), 254ff., discusses both a psychological model of divine providence, based in part on Christopher Bryant's adaption of Jungian insights to Christian theology, and a possible extension of this model to events in the empirical world, including those which are usually labelled as miraculous.

19. This means that two kinds of complexity may be defined. The first relates to the way in which, for example, evolutionary biologists do not expect the past transformation of one species into another to be repeatably demonstrable. They recognize that the complexity of the ecological niche that made this transformation possible could never be replicated. Here, the issue of repeatability is seen simply in terms of the practicability of reproducing the multifarious factors involved. The second is more conceptually subtle, and relates to holistic factors of the sort that are often seen as being involved in the emergence of new "levels of complexity" in the cosmos, such as life and intelligent self-consciousness. Here, the issue is tied to the sort of anti-reductionistic approach—advocated, for example in Paul Davies, *The Cosmic Blueprint* (London: Unwin Hyam, 1987)—in terms of new "laws" or "organizing principles" which come into operation at each emergent level in nature's hierarchy of organization and complexity. It is not necessary, Davies argues, "to suppose that these higher level organizing principles carry out their marshalling of the system's constituents by deploying mysterious new forces for the purpose, which would . . . be tantamount to vitalism . . . [instead, they] could be said to harness the existing interparticle forces, rather than supplement them, and in so doing alter the collective behaviour in a holistic fashion. Such organizing principles need therefore in no way contradict the underlying laws of physics as they apply to the constituent parts of the system" (143).

20. This point about extremity is underlined by the existence of "regime-change" phenomena such as superconductivity, which are scientifically demonstrable and understandable, but which represent discontinuities within "ordinary" experience which may have been neither predicted nor even have seemed possible before their demonstration. John Polkinghorne, *One World: The Interaction of Science and Theology* (London, SPCK, 1986), 74ff., has suggested this analogy as one that illuminates the character of events that are seen as miraculous.

21. Louth, "The Cosmic Vision of St. Maximos the Confessor"; Ware, "God Immanent yet Transcendent."

22. See, e.g., John D. Barrow and Frank J. Tipler, *The Anthropic Cosmological Principle* (Oxford: Clarendon Press, 1986), 148ff.

23. To give just two examples: the determinism of Newtonian physics has disappeared entirely through the insights of quantum mechanics, and the ontology of the world is no longer understood in terms of a naively realistic understanding of the entities posited by scientific theory.

24. Although new arguments have arisen in relation to it, the most comprehensive account of this principle is still that of Barrow and Tipler, *The Anthropic Cosmological Principle*.

25. Divine providence may be seen as "lawlike," I would argue, in the sense that identical situations will give rise to identical providential results. (This is not because God is "constrained," but because God is consistent and reliable in character.) For the reasons discussed above, however, such providence will rarely be predictable, since as we move to higher levels of complexity, situations become more difficult to identify reliably or to replicate. If we were to pursue this argument in patristic terms, then St. Augustine's implicit concept of a "higher" and a "lower" nature might prove a good starting point.

26. In particular, in Knight, *Wrestling*, I have argued for what I have called a psychological-referential model of revelatory experience that is essentially naturalistic in its framework.

27. For example, Jacques Monod, *Chance and Necessity* (London: Collins, 1972), argues that the role of chance in the universe's evolution requires an atheistic understanding. In a teleological-Christological approach, however, it is precisely the interplay of chance and necessity that allows the potential of the universe to be brought to fruition.

28. NCF, 30 above.

29. See n. 20.

30. Robert John Russell, "Bodily Resurrection, Eschatology, and Scientific Cosmology," in Ted Peters, Robert John Russell, and Michael Welker, eds., *Resurrection: Theological and Scientific Assessments* (Grand Rapids: Eerdmans, 2002), 3ff.

31. This has been more evident, perhaps, in Eastern discussion of the sacraments than in discussion of miracles. See, e.g., Philip Sherrard, "The Sacrament," in A. J. Philippou, ed., *The Orthodox Ethos: Essays in Honour of the Centenary of the Greek Orthodox Archdiocese of North and South America*, vol.1 (Oxford: Holywell Press, 1964).

Response 4. Empirical Theology and a "Naturalistic Christian Faith" / Peters

1. Arthur Peacocke, *Creation and the World of Science* (Oxford: Oxford University Press, 2004 [1979]).

2. For an excellent survey of empirical theology, see Randolph Crump Miller, ed., *Empirical Theology: A Handbook* (Birmingham, Ala.: Religious Education Press, 1992).

3. Karl Peters, "Creation and Salvation: An Attempt at Theological Construction in Relation to Evolution," in *Zygon* (in press).

4. Karl Peters, *Dancing with the Sacred: Evolution, Ecology, and God* (Harrisburg, Pa.: Trinity Press International, 2002).

5. Karl Peters, "Confessions of a Practicing Naturalistic Theist: A Response to Hardwick, Pederson, and Peterson," *Zygon* 40 (September 2005), 703–5.

6. Ibid., 705–9.

7. See Peacocke's Essay, Chap. 4.

8. Ibid., Chap. 7.

9. See Peters, *Dancing with the Sacred*, 30–37; as well as Peters, *Confessions*, 708–9. Also see Gordon Kaufman, *In Face of Mystery: A Constructive Theology* (Cambridge, Mass.: Harvard University Press, 1993).

10. Peacocke, Essay, Chapter 4.

11. Ibid.

12. *Spirit and Nature: The Moyers Collection*, Videocassette (Princeton, N.J.: Films for the Humanities & Sciences, 1997).

13. Of course, like Peacocke, Muslim thinkers also speak of the transcendent God positively.

In a lecture, Nasr combines the campfire-darkness metaphor with the infinite-ocean metaphor used by Augustine and Peacocke; see Nasr, "Islam and the Environmental Crisis," in *Spirit and Nature: Why the EnvironmentIis a Religious Issue*, eds. Steven C. Rockefeller and John C. Elder (Boston: Beacon Press, 1992), 89–90. Still, this issue remains for me. If one is an empirical theologian, one must be careful about saying things about whatever is ontologically transcendent.

14. Henry Nelson Wieman, *The Source of Human Good* (Carbondale, Ill: Southern Illinois University Press, 1946).

15. Ibid., 54-58.

16. See Kaufman, *In the Beginning . . . Creativity* (Minneapolis: Fortress Press, 2004). Although Kaufman does not call himself a naturalist, our views of God are very similar. Others consider him a possible theological naturalist; see Jerome A. Stone, "Is God Emeritus? The Idea of God among Religious Naturalists," in *Journal of Liberal Religion: An Online Theological Journal Devoted to the Study of Liberal Religion*, http://www.meadville.edu/LL_JournalLR.htm, Spring 2005.

17. See Peters, *Dancing with the Sacred*, 134–35.

18. I thank Peacocke for encouraging me to think in terms of musical analogies in the late 1970s with his book *Creation and the World of Science* (105–6). He and I use the jazz metaphor in slightly different ways. This is illustrated by my present discussion in comparison with the following. In *Theology for a Scientific Age: Being and Becoming—Natural, Divine, Human* (Minneapolis: Fortress Press, 1993), 175, he suggests that both the fugue and jazz introduce the idea of improvisation into the "model of God as composer [which] incorporates that element of open adaptability which any model of God's relation to a partly non-deterministic world should, however inadequately, represent."

19. Peacocke, Essay, Chap. 3.

20. Ibid., Chap. 5.

21. Ibid.

22. Ibid.

23. Some suffering is a necessary part of continual creation, so that nature might be called "cruciform"; see Peters, *Dancing with the Sacred*, 110–25. However, it also seems that some suffering may not be related to creating, which makes it unredeemed suffering.

24. Peacocke, Essay, Chap. 10.

25. Tyron Inbody, "History of Empirical Theology," in *Empirical Theology*, 18–19.

26. Nancy Frankenberry, "Major Themes of Empirical Theology," in *Empirical Theology*, 45.

27. Peacocke, Essay, Chap. 10.

Response 5. Sacrament and Sacrifice/ Braxton

1. E. O. James, *Sacrifice and Sacrament* (London: Thames and Hudson, 1962), 233.

2. Steven Mithen, *The Prehistory of the Mind: The Cognitive Origins of Art, Religion and Science* (London: Thames and Hudson, 1996), 51–85.

3. Terrence W. Deacon, *The Symbolic Species: The Co-Evolution of Language and the Brain* (New York: W. W. Norton and Company, 1997), 349–65.

4. James, *Sacrifice and Sacrament*, 20–25; Mithen, *Prehistory of the Mind*, 154–59.

5. Michael Shermer, *The Science of God and Evil* (New York: Times Books, 2004), 40–47.

6. Ibid, 20.

7. Eric J. Sharpe, "The Study of Religion in Historical Perspective," in *The Routledge Companion to the Study of Religion*, ed. John R. Hinnells (London: Routledge, 2005), 21–45.

8. James, *Sacrifice and Sacrament*, 233.

9. Richard Dawkins, *The Ancestor's Tale* (Boston: Mariner Books, 2004).

10. James, *Sacrifice and Sacrament*, 235.

11. Loyal Rue, *Religion Is Not about God* (New Brunswick, N.J.: Rutgers University Press, 2005), 126.

12. Ibid, 122.

13. David Abram, *The Spell of the Senuous* (New York: Vintage Books, 1996), 270–73.

14. All citations are from the transcripts for the film, *The Feeling of Science: Bridges between Science and Religion* (Huntingdon, Pa.: Juniata College, forthcoming).

Response 6. The Juxtaposition of Naturalistic and Christian Faith /Pederson

1. W. H. Auden, "For the time being. . .", Christmas Oratorio.

2. I seek this naturalistic formulation of the faith based on the Lutheran Reformation notion that *finitum capax infiniti* (the finite is the bearer of the infinite).

3. Peacocke's Essay, above, 4.

4. Ibid., 4.

5. Joseph Sittler, "The Scope of Christological Reflection" [1972], in *Evocations of Grace: Writings on Ecology, Theology and Ethics*, eds. Steven Bouma-Prediger & Peter Bakken (Grand Rapids: Eerdmans, 2000), 200–201.

6. "On this premise, one would expect the created world to reflect in its very nature the purposes of God, its Creator, and how God and God's relation to the created world are best to be articulated. Only when this foundation of insights into the nature of God and God's relation to the world has been laid should it be possible to develop an understanding of the significance of the historical Jesus of Nazareth that is, an account of Jesus as the Christ of faith." Peacocke Essay, this volume, 6.

7. Colossians 1:15-20. "And through him God was pleased to reconcile to himself all things, whether on earth or in heaven, by making peace through the blood of his cross" (New Revised Standard Version).

8. Joseph Sitter, "Essays on Nature and Grace," in *Evocations of Grace*, 118.

9. Ibid., 124.

10. Rosemary Radford Ruether, *Gaia and God: An Ecofeminist Theology of Earth Healing* (New York: Harper Collins, 1992), 235–36.

11. Veli-Matti Karkkainen, *One with God: Salvation as Deification and Justification* (Collegeville, Minn.: Liturgical Press, 2004), 26.

12. Peacocke, above, 10.

13. See an interesting new treatment of the environment and nature as it relates to the Hebrew Scriptures by Daniel Hillel, *The Natural History of the Bible: An Environmental Exploration of the Hebrew Scriptures* (New York: Columbia University Press, 2006).

14. D. Martin, *Luthers Werke: Kritische Gesamtausgabe* (Weimar: Hermann Bohlaus Nachfolger, 1938), 39, 2, 112.2–20.

15. Arthur Peacocke, *Theology for a Scientific Age* (Minneapolis: Fortress Press, 1990, 1993), 154.

16. Peacocke, *Theology for a Scientific Age*, 157.

17. Peacocke, above, 56.

18. Donna Haraway, *The Companion Species Manifesto: Dogs, People, and Significant Otherness* (Chicago, Ill.: Prickly Paradigm Press, 2003), 7.

19. Donna Haraway, *Modest_Witness@Second_Millennium. FemaleMan_Meets_OncoMouse.* (New York/London: Routledge, 1997), 42.

20. Peacocke, above, 56.

21. Haraway, *The Companion Species Manifesto*, 5.

22. Ibid.

23. Ibid., 16.

24. Ibid., 18.

25. Donna Haraway, *Modest Witness*, 134.

26. Ibid., 130.

27. Peacocke, *Theology for a Scientific Age*, 154.

28. Peacocke, above, 53.

29. Parker Palmer, *To Know as We Are Known* (San Francisco: HarperCollins, 1993), 32.

Response 7. Peacocke: Postmodern Prophet / Murphy

1. Arthur R. Peacocke, *Creation and the World of Science: The Bampton Lectures, 1978* (Oxford: Oxford University Press, 1979).

2. Bruno Latour, *We Have Never Been Modern,* trans. Catherine Porter (Cambridge, Mass.: Harvard University Press, 1993).

3. Ralph McInerny, "Introduction," in McInerny, ed., *Modernity and Religion* (Notre Dame: University of Notre Dame Press, 1994), ix.

4. Latour, *We Have Never Been Modern*, 32–33.

5. Ibid, 33.

6. Ian G. Barbour, *Issues in Science and Religion* (New York: Harper and Row, 1966), 360.

7. Peacocke, *Creation and the World of Science*, 369.

8. Stephen Toulmin, *Cosmopolis: The Hidden Agenda of Modernity* (New York: Free Press, 1990); Jeffrey Stout, *The Flight from Authority: Religion, Morality, and the Quest for Autonomy* (Notre Dame: University of Notre Dame Press, 1981).

9. Alasdair MacIntyre, *After Virtue,* 2nd ed. (Notre Dame: University of Notre Dame Press, 1984).

10. See "Plan des travaux scientifiques necessaires par organiser la societe," in *Opuscules de philosophie sociale 1819–1828* (Paris, 1883).

11. Peter Winch, "Max Weber," in Paul Edwards, ed., *Encyclopedia of Philosophy* (New York: Macmillan, 1967), 8: 280.

12. Charles Taylor, *Sources of the Self: The Making of the Modern Identity* (Cambridge, Mass.: Harvard University Press, 1989), 4.

13. Charles Taylor, *Philosophy and the Human Sciences: Philosophical Papers, Vol. 2* (Cambridge: Cambridge University Press, 1985).

14. Nancey Murphy and George F. R. Ellis, *On the Moral Nature of the Universe: Theology, Cosmology, and Ethics* (Minneapolis: Fortress Press, 1996), chap. 5.

15. Peter Berger, *Invitation to Sociology: A Humanistic Perspective* (New York: Doubleday, 1963), 69.

16. Reinhold Niebuhr, *Moral Man and Immoral Society* (New York: Charles Scribner's Sons, 1932), 257–58.

17. See, for example, Richard Dawkins, *Unweaving the Rainbow: Science, Delusion and the Appetite for Wonder* (Boston and New York: Houghton Mifflin, 1998).

18. Thomas M. Ross, "The Implicit Theology of Carl Sagan," *Pacific Theological Review* 18:3 (1985): 24–32.

19. James Turner, *Without God, without Creed: The Origins of Unbelief in America* (Baltimore and London: Johns Hopkins University Press, 1985).

20. Quoted in Merold Westphal, *Suspicion and Faith: The Religious Uses of Modern Atheism* (Grand Rapids: Eerdmans, 1993), 13.

21. Ibid, 13.

22. J. C. A. Gaskin, ed., *Varieties of Unbelief: From Epicurus to Sartre* (New York: Macmillan, 1989), 88.

Response 8. Arthur Peacocke on Method in Theology and Science / Russell

1. Arthur R. Peacocke, *Science and the Christian Experiment* (London: Oxford University Press, 1971).

2. *Idem, Creation and the World of Science: The Bampton Lectures, 1979* (Oxford: Clarendon Press, 1979).

3. *Idem, Intimations of Reality: Critical Realism in Science and Religion: The Mendenhall Lectures, 1983* (Notre Dame, Ind.: University of Notre Dame Press, 1984), 94.

4. *Idem, God and the New Biology* (San Francisco: Harper & Row, 1986).

5. Idem, *Theology for a Scientific Age: Being and Becoming—Natural, Divine and Human*, Enlarged Edition (Minneapolis: Fortress Press, 1993).

6. Idem, "Biological Evolution—A Positive Theological Appraisal," in *Evolutionary and Molecular Biology: Scientific Perspectives on Divine Action*, ed., Robert John Russell, William R. Stoeger, S.J. and Francisco J. Ayala (Vatican City State; Berkeley, California: Vatican Observatory Publications; Center for Theology and the Natural Sciences, 1998), 357–76; Arthur R. Peacocke, "The Sound of Sheer Silence: How Does God Communite with Humanity?" in *Neuroscience and the Person: Scientific Perspectives on Divine Action*, ed., Robert John Russell, et al. (Vatican City State; Berkeley, California: Vatican Observatory Publications; Center for Theology and the Natural Sciences, 1999).

7. See, for example, Arthur R. Peacocke, "Reductionism: A Review of the Epistemological Issues and Their Relevance to Biology and the Problem of Consciousness," *Zygon* 11.4 (December 1976); Arthur R. Peacocke, "Sociobiology and Its Theological Implications," *Zygon* 19 (1984): 171–84; Arthur R. Peacocke, "Science and God the Creator," *Zygon* 28.4 (December 1993); Arthur R. Peacocke, "The Religion of a Scientist: Explorations in Reality (Religio Philosophi Naturalis)," *Zygon* 29:4 (December 1994); and Arthur R. Peacocke, "Biology and a Theology of Evolution," *Zygon* 33:1 (March 1998): 695–712.

8. Arthur R. Peacocke, *An Introduction to the Physical Chemistry of Biological Organization* (Oxford: Clarendon Press, 1983).

9. I have the honor of being a member of SOSc.

10. Arthur R. Peacocke, *Theology for a Scientific Age*. See particularly Fig. 3, 217, and the accompanying text.

11. Ian G. Barbour, *Religion in an Age of Science*, Gifford Lectures; 1989–1990 (San Francisco: Harper & Row, 1990).

12. See, for example, the way John Dominic Crossan uses the word "metaphorical" as closely related to "figurative" and entirely distinct from "literal." William Lane Craig, unfortunately, seems to accept Crossan's distinction by insisting on the literal meaning of such concepts as the resurrection of Jesus. Neither appear to recognize that metaphor has a robust form in which its referential content is assumed even though it may only succeed in referring partially. See Paul Copan, ed., *Will the Real Jesus Please Stand Up? A Debate between William Lane Craig and John Dominic Crossan* (Grand Rapids: Baker Books, 1998).

13. Peacocke, *Intimations of Reality*, 41–44.

14. See the remarkable similarity between Peacocke's arguments here and those of Ian G. Barbour in Barbour, *Myths, Models, and Paradigms: A Comparative Study in Science & Religion* (New York: Harper & Row, 1974), esp. Chs. 3 and 4, written a decade before Peacocke's Lectures. See also Robert John Russell, "A Critical Appraisal of Peacocke's Thought on Religion and Science," *Religion & Intellectual Life* 2/4 (1985).

15. See Arthur R. Peacocke, "Intimations of Reality: Critical Realism in Science and Religion," *Religion & Intellectual Life* 2/4 (1985).; Peacocke, *Theology for a Scientific Age*.

16. Robert John Russell, "Entropy and Evil," *Zygon* 19:4 (December 1984): 449–68.

17. Paul Ricoeur, "Metaphor and Symbol," in *Interpretation Theory: Discourse and the Surplus of Meaning* (Fort Worth, Tex.: Texas Christian University Press, 1976), 45–69; Sallie McFague, *Metaphorical Theology: Models of God in Religious Language* (Philadelphia: Fortress Press, 1982); Sallie McFague, *Models of God: Theology for an Ecological, Nuclear Age* (Philadelphia: Fortress Press, 1987); and Sallie McFague, *The Body of God: An Ecological Theology* (Minneapolis: Fortress Press, 1993). McFague draws on both Ricoeur and Barbour (particularly *Myths, Models and Paradigms*) for her work.

18. I returned to the ambiguity of the role of thermodynamics in relation to natural good and evil in a more extended treatment of suffering in nature in Robert John Russell, "Natural Theodicy in an Evolutionary Context: The Need for an Eschatology of New Creation," in *The Task of Theology Today, V*, ed., David Neville and Bruce Barber (Hindmarsh, Australia: ATF Press, 2005), 121–52.

19. Peacocke, *Theology for a Scientific Age*, 217.

20. See Chapter 5 of the major piece by Peacocke, "A Naturalistic Faith for the Twenty-First Century: An Essay in Interpretation." See also Peacocke, *Theology for a Scientific Age*, 280–81.

21. In my view, drawn from such scholars as Gerald O'Collins, Wolfhart Pannenberg, Ted Peters, John Polkinghorne, and N. T. Wright, the resurrection of Jesus, as a proleptic act of God inaugurating the new creation at the first Easter, is more discontinuous than continuous with the world as it now is, whereas emergence refers to processes whose mild discontinuity presupposes and builds upon underlying and massive continuities in nature. See Robert John Russell, "Bodily Resurrection, Eschatology and Scientific Cosmology: The Mutual Interaction of Christian Theology and Science," in *Resurrection: Theological and Scientific Assessments*, ed., Ted Peters, Robert John Russell and Michael Welker (Grand Rapids: Eerdmans, 2002).

22. Note the difference between Peacocke's scheme and that posed by Murphy and Ellis in Nancey Murphy and George F.R. Ellis, *On the Moral Nature of the Universe: Theology, Cosmology, and Ethics*, Theology and the Sciences (Minneapolis, Minn.: Fortress Press, 1996), fig. 9.3, 204. Needless to say, I agree with Peacocke here.

23. For an initial response, see Russell, "Bodily Resurrection, Eschatology and Scientific Cosmology."

24. See Barbour, *Religion in an Age of Science*, esp. fig. 1, 32, and Fig. 2, 36. For an analysis of Barbour's analogy between theological and scientific methodologies, see Robert John Russell, "Ian Barbour's Methodological Breakthrough: Creating the 'Bridge' between Science and Theology," in *Fifty Years in Science and Religion: Ian G. Barbour and His Legacy*, Ashgate Science and Religion Series, ed., Robert John Russell (Aldershot, England: Ashgate Publishing Co., 2004), 59–76.

25. *From science to theology*: There are at least five ways or paths by which the natural sciences can affect constructive theology. (Again, I will focus on physics and cosmology for specificity, but my comments would apply to the other sciences as well.) In the first four, theories in physics, including the key empirical data they interpret, can act as *data for theology* both in a direct sense (1. and 2.) and indirectly, via philosophy, (3. and 4.). Thus:

> 1. Theories in physics can act directly as data which place constraints on theology. So, for example, a theological theory about divine action should not violate or ignore special relativity.
>
> 2. Theories in physics can act directly as data either to be 'explained' by theology or as the basis for a theological constructive argument. For example, $t=0$ in Big Bang cosmology could be explained theologically via *creatio ex nihilo*. Such an explanation can serve to confirm a theological theory, although proof is out of the question.
>
> 3. Theories in physics, after philosophical analysis, can act indirectly as data in theology. For example, an indeterministic interpretation of quantum mechanics can function within a theology of divine action by providing, after appropriate philosophical interpretation, the basis for a non-interventionist approach to divine action.
>
> 4. Theories in physics can act indirectly as the data for theology when they are incorporated into a fully-articulated philosophy of nature (e.g., that of Alfred North Whitehead).
>
> 5. Theories in physics can function heuristically in the theological context of discovery by providing conceptual inspiration, experiential inspiration, moral inspiration, or aesthetic inspiration. So biological evolution may inspire a sense of God's immanent, ongoing creativity in nature.
>
> *From theology to physics*: To see the genuinely interactive, but asymmetrical, nature of the relations I am proposing, I will suggest at least three paths by which theology can influence science. Thus:
>
> 6. Theology provides some of the philosophical assumptions which underlie scientific methodology. Historians and philosophers of science have shown in detail how the doctrine of *creatio ex nihilo* played an important role in the rise of modern science

by combining the Greek assumption of the rationality of the world with the theological assumption that the world is contingent. Together these helped give birth to the empirical method and the use of mathematics to represent natural processes. Other assumptions grounded in the *ex nihilo* tradition, however, were not carried over into the scientific conception of nature, including goodness and purpose. It would be interesting to reopen the question of the value of these assumptions for contemporary science.

7. Theological theories can act as sources of inspiration in the scientific "context of discovery," i.e., in the construction of new scientific theories. For example a variety of theologies and philosophies influenced many of the pioneers of quantum theory in the period 1900–1930, including Vedanta for Erwin Schroedinger, Baruch Spinoza for Albert Einstein, and Soren Kierkegaard for Niels Bohr. Another example is the subtle influence of atheism on Fred Hoyle's search for a "steady state" cosmology.

8. Theological theories can lead to "selection rules" within the criteria of theory choice in physics. If one considers a theological theory as true, then one can delineate what conditions must obtain within physics for the possibility of its being true. These conditions in turn can serve as reasons for an individual research scientist or group of colleagues to pursue a particular scientific theory. For example, according to theological anthropology, humankind bears the *imago dei*, which includes libertarian free will and, with it, the possibility of enacting our choices bodily. Thus we might prefer a theory that views genes as merely predisposing human behavior than one that endorses full-blown genetic determinism.

26. First, scholars in each field would need to find that such an interaction was fruitful according to the criteria of their own field of research. So, would scientists feel that their research was more fruitful by having engaged with theology and philosophy in these ways? Would theologians consider their research to have benefitted from engaging with science? Second, as major changes occur in one field and these changes are taken seriously by the other, would the corresponding effect of these changes be considered fruitful by scholars in that field? Finally, a process such as this, once set in motion, could continue indefinitely. It might even be possible to compare these results with those of scientists and theologians who have chosen not to engage in mutual interaction and determine which methodology—interaction or "two worlds" is indeed more fruitful.

27. See Chap. 4 in his Essay, "A Naturalist Christian Faith," above. Also see Peacocke's development of panentheism in the Gifford lectures and in more recent writings.

28. *Theology for a Scientific Age*, 139–40. Peacocke lays out four of the widely shared reasons for this rejection: (1) interventionism assumes that God is entirely external to the world except in special events; (2) it undermines the claim that God is the source of the rationality and regularity of the world as seen in the laws of nature, since God apparently must abrogate this regularity to act; (3) it runs aground of David Hume's arguments against evidence of the miraculous; and (4) the apparent capriciousness of special acts challenges the moral character of God.

29. Ibid., 140.

30. *Theology for a Scientific Age*, 152–57. Peacocke restates the arguments found in this section in a much clearer fashion in "God's Interaction with the World" found in *Chaos and Complexity*, 276, n. 32. I will follow the latter presentation, but provide page references to both sources so the reader may compare them independently.

31. "God's Interaction," 266–72; *Theology for a Scientific Age*, 44–55, 152–55.

32. *Theology for a Scientific Age*, 153; "God's Interaction," 277.

33. "God's Interaction," 278. Note his criticisms of W. Alston and others in n. 34. *Theology for a Scientific Age*, 154. In passing he levels this criticism at John Polkinghorne's arguments about divine action and nonlinear dynamics.

34. "God's Interaction," 279.

35. Ibid., 280. *Theology for a Scientific Age*, 155.

36. Ibid., 281.

37. See his Essay, "A Naturalist Christian Faith," above, Chap. 9. For a more detailed account see *Theology for a Scientific Age*, 158–59, including the wonderful quotation from St. Augustine.

38. My own preference for their combination is what can be achieved through a more explicitly Trinitarian conception of divine agency.

39. *Theology for a Scientific Age*, 158. See also *Creation and the World of Science* for an earlier discussion of panentheism (45, 141, 201, 207, 352). In *Theology*, n. 75 (370–72), Peacocke includes a very helpful discussion on the definition of panentheism, its relative independence from any specific metaphysical system (*pace* process theologians and John Polkinghorne!), a note about his (I think very wise) intention to avoid using the term because of its frequent misinterpretations, and some comments on the way it is formulated by Charles Hartshorne and others.

40. In "God's Interaction" he prefers the term "whole-part constraint" (282–83).

41. There are, of course, rich philosophical resources for deploying the concept of the "world-as-a-whole," but in general Peacocke tends not to use them. See, for example, his opting out of a close connection with process metaphysics as mentioned in a previous endnote. Whether a more developed use of such resources would actually help him take advantage of the scientific view of the world is another question.

42. See Peacocke, *Theology for a Scientific Age*, 194-95, fig. 1; *idem*, "The Sound of Sheer Silence: How Does God Communicate with Humanity?" in *Neuroscience and the Person: Scientific Perspectives on Divine Action*, ed., Robert John Russell et al. (Vatican City State; Berkeley, California: Vatican Observatory Publications; Center for Theology and the Natural Sciences, 1999), 238, fig. 1.

43. The "*q*" does not appear in the 1993 version of the diagram. Peacocke added it to the 1999 version following the suggestion I made that the figure needed something to indicate divine immanence along with divine transcendence.

44. *Theology for a Scientific Age*, 192. See also "The Sound of Sheer Silence," 235.

45. Indeed one can formulate general relativity, on which contemporary cosmology is based, and distinguish it from other twentieth-century theories of gravity, by the aphorism "the boundary of the boundary is zero;" written topologically, $\delta\delta = 0$. Astonishingly, Einstein's field equations, $G_{;<} = 8BT_{;<}$, link the geometry of spacetime $(G_{;<})$ with the distribution of stress-energy $(T_{;<})$ such that the condition $\delta\delta = 0$ on spacetime geometry implies the conservation of energy-momentum. Hence one need not impose it as a separate condition as in other competing theories of gravity in the twentieth century. All the more reason to conceive of the universe as "boundaryless." See Charles W. Misner, Kip S. Thorne, and John Archibald Wheeler, *Gravitation* (San Francisco: W. H. Freeman & Co., 1973), Chap. 15.

46. Even in more complicated models, such as inflationary Big Bang and recent speculations in quantum gravity, the universe is unbounded.

Response 10. On Divine and Human Agency / Clayton

1. Alfred North Whitehead, *Science and the Modern World*, Lowell Lectures 1925 (New York: Macmillan, 1925), 275.

2. Whitehead, *Science and the Modern World*, 270.

3. See for example Stuart Kauffman, "Beyond Reductionism: Reinventing the Sacred," *Zygon* (forthcoming, 2007).

4. Thomas Aquinas, *Summa contra Gentiles*, III, Q. 103.

5. C. S. Lewis, *Miracles: A Preliminary Study* (New York: Macmillan, 1977 [1947]).

6. See Maurice Wiles, "Religious Authority and Divine Action," in *God's Activity in the World: The Contemporary Problem*, ed., Owen C. Thomas (Chico, Calif.: Scholars Press, 1983), 186.

7. See Wiles, *God's Action in the World* (London: SCM Press, 1993), 108.

8. See Gordon Kaufman, "On the Meaning of Act of God," *Harvard Theological Review* 61 (1968): 175–201. For a helpful presentation of these authors see Nicholas Saunders, *Divine Action and Modern Science* (Cambridge: Cambridge University Press, 2002), Chap. 2.

9. See James Kellenberger, "Miracles," *International Journal for the Philosophy of Religion* 10 (1979): 145–62, quote 157.

10. Saunders, *Divine Action,* 51.

11. Arthur Peacocke, *God and Science: A Quest for Christian Credibility* (London: SCM Press, 1996), 19.

12. See Peacocke, "Emergence, Mind, and Divine Action: The Hierarchy of the Sciences in Relation to the Human Mind-Brain-Body," in *The Re-Emergence of Emergence*, eds., Philip Clayton and Paul Davies (Oxford: Oxford University Press, 2006), 267 n.10.

13. Ibid., 265, emphasis added.

14. Ibid., 264.

15. Ibid., 269.

16. Ibid., 270.

17. Ibid., 271.

18. Ibid., 267 n.11.

19. Ibid., 271.

20. Ibid., 272.

21. Ibid., 271, 273.

22. I examine the correlation in chap. 6 of *In Quest of Freedom: The Emergence of Spirit in the Natural World* (Göttingen: Vandenhoeck & Ruprecht, forthcoming 2007).

23. See Philip Clayton, *God and Contemporary Science* (Grand Rapids: Eerdmans, 1997), chapter 8.

24. An indirect influence might still be possible. For example, God might bring about direct changes at the micro-physical level, which might then be augmented by some mechanism until they induced changes in human thought.

25. Peacocke, "Emergence, Mind, and Divine Action," and "Emergent Realities with Causal Efficacy—Some Philosophical and Theological Applications," in *Evolution and Emergence: Systems, Organisms*, eds. Nancey Murphy and William R. Stoeger, S.J. (Oxford: Oxford University Press, forthcoming 2007).

26. See Clayton and Peacocke, eds., *In Whom We Live and Move and Have Our Being: Panentheistic Reflections on God's Presence in a Scientific World* (Grand Rapids: Eerdmans, 2004).

27. See Peacocke, *Paths from Science towards God: The End of All Our Exploring* (Oxford: One World Publications, 2001), 114.

28. This "only if" phrase demands one qualification. Throughout this response I assume—as I think Peacocke does also—that one needs to give some sort of an account of what one means by divine action language. This claim may be, and often is, disputed by authors on this topic. Thus some have argued that no conceptual account is necessary because all language about God is symbolic, apophatic, regulative, pragmatically useful, or "internal to the practice of faith." Any one of these approaches might allow one to speak of divine "acts" in the world (the scare quotes now become crucial!) or to label various events as "expressions of divine grace."

29. "Emergence, Mind, and Divine Action," 273–74.

30. Peacocke, "Emergent Realities with Causal Efficacy," MS 19.

31. "Emergence, Mind, and Divine Action," 273.

32. Ibid., 273–74.

33. Many aspects of the interpretation and constructive position presented here were developed in multiyear correspondence and conversations with Steven Knapp, whom I also thank for detailed criticisms of an early draft of this essay.

Reflections on the Responses / Peacocke

1. In "Science and the Future of Theology—Some Critical Issues," *Zygon* 35 (2000): 119–40.

2. For references see the recent publication of the significantly entitled *The Re-emergence of Emergence: The Emergentist Hypothesis from Science to Religion,* ed. Philip Clayton and Paul Davies (New York: Oxford University Press, 2006).

3. For example, Drees questions those who take the early Jesus tradition as normative.

4. See, for example, my *Theology for a Scientific Age* (2nd enlarged edition, Fortress Press, Minneapolis and SCM Press, London), 279–88.

5. By "spiritual," I here refer, in this context only, to any experiences by human beings of nature and humanity that have to rely on language that refers to entities, structures, and processes deployed in the religious traditions.

6. To use my oft-repeated term.

7. Or whatever is appropriate in the language of the participant.

8. To follow the usage in my Essay.

9. Philip Clayton refers to this analysis of mine in his response in this volume (171). I have developed this analysis in "Emergence, Mind, and Divine Action: The Hierarchy of the Sciences in Relation to the Human: Mind-Brain-Body," in Clayton and Davied, eds., *The Re-Emergence of Emergence*; and in "Emergent Realities with Causal Efficacy—Philosophical and Theological Applications," in *Evolution and Emergence: Systems, Organisms, Persons*, eds. N. Murphy and W. R. Stoeger, S.J. (Oxford: Oxford University Press, forthcoming 2007).

10. Serendipidity is "the faculty of making happy and unexpected discoveries by accident" (*Oxford English Dictionary*).

11. *Zygon* 41/ 2 (June 2006), 347–63.

12. Complementary to P. Hefner's well-known "created co-creators."

13. Donna Haraway is not a theist, Pederson reports.

14. "Inference to the best explanation" also characterizes this position, and I am far from repudiating the Enlightenment.

15. He refers to my St. Andrews Gifford Lectures as presented in my *Theology for a Scientific Age* (1993). They were also presented in my *God and Science* (London: SCM Press, 1996).

16. Russell even goes so far as to say that he is in "solid theological agreement" with me in my "understanding of God's relation to the world"—meaning my emphasis on "God not only sustaining the world but interacting continually with it to achieve both general and particular effects" (148). He approves of my speaking here, in this context, of the "world-as-a-whole."

17. According to Boethius, God's eternity is the "total, complete, simultaneous possession of eternal life."

18. [This quotation appears in an earlier draft of Professor Ward's response but not in the published version. – Ed.]

19. For an account of the various levels in both nature and in human persons see my *Theology for a Scientific Age*, chap. 12.

20. For a fuller account of my analysis of the New Testament miracles in particular, see *Theology for a Scientific Age*, 274–88.

21. As published in the series subtitled, "Scientific Perspectives on Divine Action" edited by R. J. Russell, et al. (Vatican City State: Vatican Observatory Publications; Berkeley, Calif.: Center for Theology and the Natural Sciences).

22. Drawing upon recent papers of mine—for example, his references 11 and 18.